IN THE BEAT OF A HEART

IN THE BEAT OF A HEART

LIFE, ENERGY, AND THE UNITY OF NATURE

JOHN WHITFIELD

Joseph Henry Press
Washington, D.C.

Joseph Henry Press • 500 Fifth Street, N.W. • Washington, DC 20001

The Joseph Henry Press, an imprint of the National Academies Press, was created with the goal of making books on science, technology, and health more widely available to professionals and the public. Joseph Henry was one of the founders of the National Academy of Sciences and a leader in early American science.

Library of Congress Cataloging-in-Publication Data

Whitfield, John, 1970-
  In the beat of a heart : life, energy, and the unity of nature / John Whitfield.
    p. cm.
  Includes bibliographical references and index.
  ISBN 0-309-09681-2 (hardcover)
  1. Life (Biology) 2. Vital force. 3. Nature. I. Title.
  QH501.W48 2006
  570—dc22

                              2006012277

Cover design by Michele de la Menardiere.

Cover credit: Mouse photo © GK Hart/Vikki Hart/Getty Images. Background image © Photo Researchers.

Printed in the United States of America

for Sara

CONTENTS

1   PROLOGUE: "I HAVE TAKEN TO MATHEMATICS"

As HE ENTERED his forties, in the first decade of the twentieth century, D'Arcy Wentworth Thompson looked back on his life with some bitterness. He was stagnating, teaching zoology at a provincial university in a coarse, industrial Scottish city. His fellow scientists, when they thought of him at all, seemed to view him as a dilettante and showed no interest either in his ideas or in giving him a better job. He, in turn, did not think much of their politicking and lazy orthodoxy—"the brethren," he called them. In his bottom drawer was a translation of Aristotle's work on animals, still unfinished 30 years after it was begun.

Fortunately, Thompson had something to take his mind off his frustrations. For the past 20 years, besides doing the everyday, unspectacular work of the Edwardian biology professor—describing, counting, and cataloguing living specimens—he had quietly been building a new foundation for his science from the techniques and concepts of physics and mathematics. These other fields, he believed, could describe and explain how living things were constructed and how they functioned.

This was radical work. Many scientists of the time believed that living things were fundamentally different from any chemical or physical system and beyond the powers of physics or chemistry to explain. This school of thought held that living matter was imbued with unique properties or animated by a special force—a philosophy called vitalism. To vitalists, biology's task was to seek out the differences between living and dead things, not the commonalities. Such a view was not merely the territorial behavior or professional blinkeredness of biologists. Even eminent physicists such as James Clerk Maxwell, whose work revolutionized our understanding of electromagnetism and gases, believed that the chemistry of atoms, molecules, and elements could not explain living matter and that the laws of thermodynamics could not explain how bodies worked. And few biologists knew any mathematics, partly because it seemed irrelevant to their discipline and partly because of a feeling that to explain living forms in this way, to show that they fell within the orbit of sciences developed to explain dead matter, belittled them in a way that was almost immoral.

But Thompson was different. His familiarity with, and love of, the works of Aristotle, Pythagoras, and Plato had given him an expertise in mathematics, particularly geometry, far beyond that of most zoologists. He was a voracious polymath, willing to take inspiration anywhere. Classical, or renaissance, or enlightenment thinking was not obsolete; it was fundamental: "A fact discovered yesterday is balanced by the history of two thousand years," he wrote. His intellectual roots lay in the time when educated people read and pondered everything, seeing no distinction between different branches of scholarship. And instead of seeing it as a debasement to bring mathematics into biology, he had the Greek belief in that discipline's beauty and perfection.

Another thing that made Thompson an unusual biologist is that he thought his science should, following the example of physics, look for big answers and grand, overarching theories. Faced with the dazzling complexity of life, biology has traditionally been a science built of details and descriptions—from Linnaeus's classification of species to the modern drive to sequence genomes. But Thompson belonged to another strand, which asks: Is there an underlying unity to nature? Can we discern patterns running through life's diversity? Are there rules that might explain these patterns? And if so, are these rules the product of chance and history, or can we perceive them as the inevitable product of constant forces and conditions? And can these forces and conditions be expressed in the language of mathematics, physics, and chemistry?

In 1917, Thompson published his thoughts on these matters in a book, *On Growth and Form*. The work brought him intellectual fulfillment and professional recognition, and it has become a scientific and literary classic, second only to Darwin's works in the pantheon of biological writings. But it was a beginning, not an end—"all preface," said Thompson. We have yet to reach the end: In the intervening century, his preoccupations have never occupied center stage in biology, but neither have they gone away. His questions still need answers.

This is the story—with some detours—of D'Arcy Thompson's strand of biology and of a century-long attempt to build a unified theory, based on the laws of physics and mathematics, of how living things work. At the story's heart is the study of something that Thompson called "a great theme"—the role of energy in life. "Morphology," he wrote, "is not only a study of material things and of the forms of material things, but it has its dynamical aspect, under which we deal with the interpretation . . . of the operations of Energy." Energy is life, and life's currency. It unites and divides all living things; its flow from one place to another controls everything from cells to forests. If you follow it, you can understand how nature works. As the physicist Ludwig Boltzmann wrote in 1886: "Available energy is the main object at stake in the struggle for existence and the evolution of the world."

The way that energy affects life depends on the size of living things. Size is the most important single notion in our attempt to understand energy's role in nature. Here, again, we shall be following Thompson's example. After its introduction, *On Growth and Form* ushers the reader into a physical view of living things with a chapter titled "On Magnitude," which looks at the effects of body size on biology, a field called biological scaling. Growth and form, Thompson explains, are both questions of body size. The former represents changes in size with time, and "the form of an object is defined when we know its magnitude."

We are going to see how far an understanding of biological energy, viewed through the lens of body size, will get us in understanding life, from how much food we need, to how long we live and how many offspring we can be expected to produce, to the way that species are spread across the globe and how tropical forests get to be so exciting. D'Arcy Thompson will point the way, as both figurehead and navigator.

Father and Son

D'Arcy Wentworth Thompson took his name from his father. His father took his name from Captain D'Arcy Wentworth of the British army who, in April 1829, traveled on the sailing barque *Georgiana* from the Australian penal colony of Van Diemen's Land to Sydney. A lack of other work had driven the ship's master, John Skelton Thompson, to take a job transporting British convicts to Australia. The *Georgiana's* journey to the southern hemisphere had been hard, beset by storms, sickness, and death among the prisoners and knife fights among the crew. Thompson had an additional worry: His new and pregnant bride, Mary, was aboard—the ship was their only home. She gave birth to a son on April 17, with the ship in sight of its destination, where it would soon unload its troublesome cargo.

As a young man of good character, Captain Wentworth offered himself as the child's godfather, and so also gave D'Arcy Wentworth Thompson his name. Mother and son arrived back in Europe two years later, settling in Brussels. John Thompson did not, having died of sunstroke in the West Indies. The family was not well off, but the boy won a scholarship to Christ's Hospital School in England, where he spent what he called a "dreary, weary boyhood," not seeing his mother for 12 years. At school Thompson excelled in the classics. On holidays he walked the countryside alone and learned to imitate bird calls. At the age of 19, he won another scholarship, allowing him to continue his classical studies at Trinity College in Cambridge. After graduating, he applied for a fellowship at Trinity but was turned down, reputedly because he had once attended the college's chapel in his dressing gown. Instead, he got a job as a classics master at the Edinburgh Academy, where his pupils included Robert Louis Stevenson, and his imaginative teaching, rejection of corporal punishment, and advocacy of higher education for women won him renown and opprobrium in roughly equal quantities.

In 1858, Thompson met and fell in love with Fanny Gamgee. The Gamgees were a prominent Edinburgh scientific family, and Fanny's father Joseph and her brother John both held lectureships in veterinary medicine. Like the Thompsons, the Gamgees were self-made and well traveled. Joseph, an influential horse doctor, had learned his skills in Europe, where he served an Italian prince before returning to study in London; to save money for his fees, he made the journey on foot. D'Arcy and Fanny were married in 1859, and Fanny gave birth to a son on May 2 of the following year. A week later she died of puerperal fever, at age 21.

The loss of his wife was a hammer blow to the husband's mind and health. He was unable to care for his newborn son, and so the baby was sent to Fanny's parents, to be raised by a maiden aunt, Clementina. Fanny's mother named the boy after his father. Thompson senior recovered and emigrated to Galway, Ireland, in 1863, where he settled, began teaching again, and eventually remarried. He left his son behind, although the two developed a deep bond, writing often and visiting occasionally. Thompson junior inherited his father's interests in the classics and natural history, spending his weekends identifying the local plants and fossils, and fishing in rock pools. The Edinburgh Academy— the son following as a pupil where his father had taught—took care of the book learning, covering the classics, the Bible, mathematics, and languages. In 1877, Thompson was top in his class in the final exam. He made up for the curriculum's lack of science by compiling a list of all the species of plants and animals he could find on the school grounds. He inherited his physique from the Thompson family's Scandinavian ancestors, growing to over 6 feet tall, with red-blond hair and piercing blue eyes.

In 1878, the younger D'Arcy Wentworth Thompson went to Edinburgh University with the intention of studying medicine. As his education progressed, however, he became more interested in zoology, botany, paleontology, and the debates around evolution that were wracking the university. He published his first scientific paper, on a fossil seal, during this time. But the notion that biology was a discipline in its own right was new and studying it involved switching universities, to Cambridge, which had appointed its first professor in the subject in 1866. Grandfather Joseph warned him not to be seduced by specialization: "The new fangled idea of subjects being so great that only parts must be undertaken by one man is a consummate absurdity." Thompson got the money to attend Cambridge by winning a scholarship, but it wasn't enough to live on, so he taught Greek and wrote for encyclopedias. Some of this work brought academic as well as financial benefit: One of his first significant scientific endeavors was to translate a German book on pollination. Charles Darwin—who had also studied medicine at Edinburgh before moving to Cambridge—wrote an admiring preface, but died before the book was published in 1883. Thompson had visited Germany in the summer of 1879 to learn the language—an essential skill for any zoologist, as the country was the center of biological science.

The biographies of Victorian scholars often make it seem as if people had more time, or energy, or talent back then. Besides making money and pursuing his studies, it was in Cambridge that Thompson embarked on the translation, in collaboration with his father, of Aristotle's *Historium Animalium*. He was also active in university life, becoming a leading light of the university's debating society. Despite all this, he seems never to have fit in, either at school or university. A contemporary at the former remembered him as "a queer fellow—there was always something about him we couldn't understand," and a professor at Cambridge once told him, "You know you haven't got many friends." Thompson was shocked, but on reflection saw the truth in this, and the feeling of friendlessness, of being apart from other people, remained with him throughout his life. But he also wore his loneliness as a badge of intellectual integrity, a pride enhanced by a belief that he had inherited the trait from his father. "He never hunted with the pack, nor barked, growled, yelped with them either," he wrote. "And, thank God, no more do I!"

Thompson also inherited his father's attitude toward work. Thompson senior believed it was impossible to reconcile ambition with integrity and that pushing oneself forward was, "[something] that a gentleman, a pukka Sahib, does not do." Likewise, his son disdained the maneuvering that an academic career demands—being nice to the right people, publishing regularly on the right subjects—and often proclaimed his contempt for careerism. And while he could be charming, devoted, and loyal, he could also be inscrutable, volatile, impulsive, and brusque, particularly to fellow scientists. Such aloofness hardly enhanced his employability. The Cambridge higher-ups thought he spread himself too thinly, starting more projects than he finished and never knuckling down to the one solid piece of work that would have made his scientific name. In three successive years, Thompson applied for fellowships at Trinity College that would have allowed him to continue studying at Cambridge, but like his father he was never successful. Despite his loner's sensibilities, Thompson craved the scientific establishment's recognition, and its failure to arrive hurt him. "I came to [Trinity] in exultation," he wrote toward the end of his life. "I left it in deepest disappointment."

A Lean and Hungry Town

Thompson may have felt snubbed at Cambridge, but his professors gave him glowing testimonials. He also got a recommendation from his uncle, Arthur Gamgee, sometimes credited as the first biochemist for his work on the chemical composition of the body and the chemical changes that accompany disease. Armed with these references, Thompson got a job back in Scotland in 1884, at the newly established University College of Dundee. Dundee had grown rich on the textiles industry, but had also grown grim—it had more money than class, and more slums than money: "An east-windy . . . lean and hungry town," Thompson wrote. It was soon to get much leaner, as the textile industry collapsed in the face of competition from India. But Dundee also had a zeal for self-improvement, and so in 1883 the college was founded, with an emphasis on the new sciences and a faculty of bright young professors to teach them. By now his father's ideas about education were more widely accepted, and Thompson junior lectured to equal numbers of women and men.

Thompson threw himself into Dundee life, getting involved in the college's administration and politics, setting up its zoology museum, and doing philanthropic work in the town. Thompson's speciality was marine biology—he loved the sea and was proud of his family's

D'Arcy Wentworth Thompson (1860–1948).
Credit: University of Dundee.

nautical heritage. He made regular visits to Dundee's docks in search of specimens fresh from fishing boats and whaling ships, and built up an outstanding collection of arctic specimens in the college museum. In 1896 this work attracted the government's attention. For 50 years the powers with interests in the North Pacific—the United States, Russia, and Great Britain, via her colony Canada—had wrangled over the regulation of the fur seal hunt. The same issues that exercise managers of commercially exploited species today were just as contentious then: how many animals could be taken, when and where, and the extent of poaching. The British foreign office commissioned Thompson to study the seal population to assess whether the hunt was, in today's jargon, sustainable. Besides his marine expertise, he had made a good impression on the foreign secretary's nephew while at Cambridge.

Thompson sailed to New York in May and traveled overland to Seattle, where he met up with an American team charged with the same mission. They boarded a boat north to the Aleutian Islands, which stretch out into the Pacific off Alaska. Thompson collected specimens of every species that crossed his path, to swell the collection in Dundee. Once he reached the seal colonies, he did all the things that a good conservation biologist should: counting the population and trying to estimate the natural levels and causes of pup mortality.

He also saw seals driven in their hundreds from breeding rookeries to killing grounds. The drive sounds a strange sight, but Thompson's account of it is unperturbed:

> The seals certainly puffed and blew and sweated and steamed; they stopped every now and then to rest . . . but after a moment they went on briskly. The signs of distress were less painful than I have often witnessed in a flock of sheep on a hot and dusty road, and I have seen drovers show less regard for the comfort of their sheep.

The drive began at 2:00 in the morning, and by 6:00, a herd of nearly 2,000 seals had reached the killing grounds, where teams of men with clubs waited.

> The men employed were clean, skilful and vigorous. A single blow, or two at most, dispatched each seal, and I saw no failure of aim, even in the confused mass of seals tumbled pell-mell over one another. They showed no signs of terror; the survivors of each batch [only about half of the driven seals were killed: those either too young or too old were released] made

quickly for the water, and were already swimming homeward as the next batch were being slain. . . . Two younger lads went round plunging a knife into the heart of any seal that still breathed, five (rippers) proceeded to slit the skins down the belly and around the neck and paws, after which the rest flayed the carcasses. The work of skinning nearly kept pace with that of killing. I could not detect in the whole process either intentional or accidental cruelty. . . . The killing was concluded by about 10 o'clock, an interval for breakfast intervening.

The next year he went to Alaska again, this time going east via Ceylon and Japan. In his final report to the Bering Sea Commission, Thompson concluded that the seal hunt was sustainable at its present level but that there was no room for complacency: "We may hope for a perpetuation of the present numbers; we cannot count on an increase." He also noted that the killing of other marine mammals—whales, walruses, and sea otters—was less well regulated and posed a greater threat to those species' survival.

The experience gave Thompson a taste for fisheries research, and for international collaborations. In 1898 he joined the fisheries board of Scotland, which was to become a lifelong commitment, and from 1902 onward he represented Scotland on the newly formed International Council for the Exploration of the Sea, lobbying the British government to support the fledgling body. All the while he kept up his classical studies. In 1901 he married Maureen Drury, his step-mother's niece. Over the next nine years, his three daughters Ruth, Molly, and Barbara were born.

But this life left Thompson little time for his own research, and his work on applied problems cut little ice with the people who gave out jobs. "You must show that the waters of the Pacific have not washed the science out of your composition," wrote an old Cambridge friend. Through the 1890s, Thompson applied for positions at more eminent institutions in London, Edinburgh, and Glasgow, but failed to get any of them. On the other hand, he applied for and won a professorship in Aberdeen but turned it down; he clearly wanted to work on his own terms, even if it hindered his progress up the greasy pole of academia. Other researchers saw him as an unproductive oddball in two fields; his papers on classical scholarship were rejected just as often as his biology ones.

Besides his political failings, Thompson slowed his advancement by taking a near-heretical view of biological science. In the early twentieth century, biologists, reveling in—or reeling from—the triumph of Darwinism, tended to use evolution as a catch-all answer to biology's "why?" questions. Why was any animal the way it was? Because natural selection favored it—never mind how. Biology was concerned with stories, not causes: Animals were seen as archives of evolution, and biologists' main task was to use specimens' anatomy to work out their history. Then, by comparison with other species, researchers could assign each creature its place on life's family tree. As the British immunologist Peter Medawar later wrote, biologists "accepted the contemporary and far from adequate form of Darwinism in much the way that nicely brought up people accept their religion, that is, in a manner that contrives to be both tenacious and perfunctory." Most zoologists were not much interested in how animals got to be the way they were, in either the long-term evolutionary sense or the short-term developmental sense. Nor were they interested in how animals worked; rather than ask why a bone took a particular form, they just wanted to compare it with the same bone in different species. But most of the striking advances in comparative anatomy were already decades old. Evolutionary biology was snoozing.

Perhaps this is why Thompson came to take a dim view of natural selection and genetics. He was interested in causes, thinking that when we see some feature in an organism, we should seek some underlying force that has brought the trait about. Simply to invoke heredity, as many biologists then did—for example, in the popular idea known as "ontogeny recapitulates phylogeny," which hypothesized that the process of an animal's embryological development re-enacted its evolutionary ancestry—seemed uselessly vague. "I should be [equally] at a loss to disprove the hypothesis that the Moon was made of green cheese," Thompson wrote of the notion that evolution could explain how living things arrived at their present forms. He was no creationist: He granted that natural selection could weed out the unfit, but doubted its power as a creative force, able to explain why life took one form and not another. He also thought that physical similarities were a poor guide to shared ancestry—another position unlikely to endear him to his peers, since that is what they spent much of their time doing. The disenchantment began in Cambridge, where he could never summon up much interest in the lineages or relationships of species. His professors advised him to keep his thoughts on evolution to himself: "You might think these things, but you mustn't say them," one warned.

Isolated in Dundee, however, Thompson could give his originality its head and, untroubled by career success, work on what he liked. In 1894 he presented a paper to the British Association meeting in Oxford titled "Some Difficulties of Darwinism," in which he expressed doubt that, among other things, the vibrant colors of hummingbirds could come about through a dour struggle for existence. Thompson, in contrast, saw the plumage as the product of a benign environment. The hummingbird's livery, he suggested, was the product of "laws of growth," operating unchecked by natural selection. He also sought to banish natural selection from the shape of birds' eggs, arguing that the narrow, pointed eggs of the guillemot, a seabird that nests on cliff ledges, were formed by pressure from muscles in the mother's oviduct. This contrasted with, although it does not contradict, the evolutionary explanation that this shape gives the eggs a tight turning circle and stops them from rolling off their ledge. One of Thompson's old Cambridge professors was in the chair. "He did not hide his impatience and disapproval," Thompson later recalled. "There were *no* difficulties in Darwinism, either to him or any sensible man in those days." In truth, Thompson's notions about evolution and natural selection have not aged well. But his efforts to formulate new laws to put in the place of natural selection took him down many fruitful paths. In the process he pioneered a different type of biological thinking and a new type of biological explanation.

Unsuspected Wonders

The seeds of mathematical and physical thinking germinated in Thompson's mind in the late 1880s. In October 1889 he wrote to Mary Lily Walker, a former student: "I have taken to Mathematics, and believe I have discovered some unsuspected wonders in regard to the Spirals of the Foraminifera!"

Foraminifera are microscopic sea creatures with intricate and beautiful shells. Thompson did not say what he had discovered, but I suspect he had noticed that many foraminifera have shells that take the form of an equiangular, or logarithmic, spiral. In this shape each whorl is broader than its predecessor, but the ratio of breadths remains constant. In other words, the shell is a coiled cone. Traveling outward from the cone's apex at the center of the shell, the tube becomes progressively broader, but always at the same rate. This spiral was to lie at the heart of Thompson's thinking for the rest of his life. Into his seventies and eighties, he was still writing to biologists and engineers, seeking to understand how the shells of sea snails and the horns of antelopes and narwhals, which are also equiangular spirals, came about.

Thompson began to investigate other marine creatures and found that mathematics could describe their form too. In some sponges, small bony particles called spicules make a scaffold for the animal's soft tissue. Thompson began to think that the form of these spicules might be determined by the animal laying down mineral crystals in gaps between its cells. Three spherical cells in contact would, like soap bubbles, produce a three-rayed crystal with equal angles of 120° between its arms, like the center of the Mercedes-Benz logo. In 1890 he shared these observations with a student, George Petrie, who was astounded: "I became aware that mathematics may be applied to give precision to biological observations and thus to open up a fascinating vista of speculations," he wrote. Physical forces gave Thompson an alternative to heredity as an explanation of the form of animals. He decided that physics was all that was needed to explain spicules' shape, and that similarities between species probably reflected shared physical conditions, not shared ancestry.

In 1894, the same year he told the British Association meeting of his concerns over Darwinism, his speculations about laws of growth, and his belief in the power of physical forces to shape living forms, Thompson wrote to Michael Foster, a comparative physiologist who had taught him at Cambridge, about this line of reasoning. Foster, one of the most powerful zoologists in Britain, was less than enthused. "I confess I am not very much attracted by this line of work, and doubt if it's likely to be fruitful," he replied. "If the form is constant in a group—

it does not matter how the form is brought about." It seems bizarre now, but Foster was only interested in using the shape of sponges' spicules as a tool for recognizing species and assigning them to different groups. He seems to have seen any investigation of how the shape of a spicule came about, or why, as irrelevant and possibly dangerous. "Does your result wholly destroy the diagnostic value of the spicules?" he wrote. Thompson published nothing else in this line until a paper in the scientific journal *Nature* in 1908, which revisited his thoughts on the shapes of eggs. In Dundee he discussed what he called his heresies with assistants but was unwilling to publish, saying "everyone will say they have read it all before." Maybe he also feared further ostracism.

Aristotle's Disciple

Yet as he entered his sixth decade, Thompson's career took flight. In 1910, eight years after his father's death, he finally completed the translation of *Historium Animalium* that the two had begun decades earlier. It remains the standard version of the work and will always be so, as there is unlikely to be another naturalist who knows as much Greek nor a classicist as expert in zoology. Thompson thought Aristotle the greatest of all biologists: "No man has ever looked upon our science with a more far-seeing and comprehending eye," he told the British Association a year later. In 1911, Cambridge awarded him a doctorate, in letters rather than science, for the Aristotle translation and another work published 15 years earlier, *The Glossary of Greek Birds*, a gazetteer of matters ornithological in Greek literature that Thompson called "the apple of my eye."

D'Arcy Thompson's later achievements make me wonder whether his earlier outsider status was largely self-imposed and that at some point he simply decided to stop shunning the brethren and embrace them. In his fifties he also began to pursue his thoughts on physics, mathematics, and biology in earnest. The aforementioned 1911 talk at the British Association's meeting in Portsmouth was delivered from his position as president of the association's zoological section. After the encomium to Aristotle, Thompson went on to outline his view of "the greater problems of biology."

The foremost of these problems was vitalism. "The hypothesis of a Vital Principle, or vital element," Thompson told his audience, "[is] the greatest question for the biologist of all." The debate had been given new form by studies of cells, embryology, and reproduction—processes that seemed almost magical, and certainly far removed from physics and chemistry. Yet just as a cellular view of life had revealed that zoologists, botanists, and physiologists were all studying essentially the same thing, Thompson argued that still deeper investigation would dissolve the distinction between biology, chemistry, and physics. Only then would we have a true understanding of biology.

Thompson thought that some sciences were more scientific than others. Chemistry outranked biology and was in turn topped by physics, with mathematics standing at the pinnacle. The problem with vitalism—and heredity—was the belief that the explanations for biological phenomena could be found in biology. Instead, biologists should go up the chain of explanations, to chemistry and beyond. They had neglected the physical sciences to their detriment. Why invoke vitalism when so many of the forms in the living world can be explained by simple physical principles? The physics of surface tension, Thompson noted, explain why raindrops are spherical, because this shape has the minimal surface area. Likewise, he argued, surface tension could explain the shape of amoeboid cells, or the spread of sticky droplets over a spider's web. "Has the biologist," Thompson asked, "fully recognized . . . that the physicist may, and must, be his guide and teacher in many matters regarding organic form? . . . In many of the simpler cases the facts are so well explained by surface tension, that it is difficult to find a place for a conflicting, much less an overriding, force." Vitalism, in short, was unnecessary.

Around this time, Thompson agreed to write up his thesis in what he called a "little one-shilling or two-shillings-and-sixpence book for the Cambridge University Press on 'Form and Growth.'" But the work grew, and any prospect of publication disappeared into the distance, as the edifice that had begun as a simple chapel grew into a cathedral. Thompson hinted at what he was up to in December 1914, in a paper presented to the Royal Society of Edinburgh titled "Morphology and Mathematics." Again, he began by blowing the trumpet for bringing biology under physics' umbrella. Biological forms, he argued, were a subset of those seen in nature, which were themselves a subset of all those theoretically—that is, mathematically—possible. Biologists should stop worrying about how a natural form, such as a shell, differed from an abstract spiral and instead learn from the similarities. Then they would see how mathematics would lead them from description, to analysis, to generalities.

But, he conceded, most of the forms in living organisms were far too complex to yield their secrets to such simple analytical tools. This, however, did not exclude the use of mathematics to compare different shapes. Adapting a technique developed by renaissance artists, particularly Albrecht Dürer, he laid coordinate grids over pictures of animals, or shells, or bones and showed how distorting them in a regular fashion could produce the forms of other related species, like the same landmass shown in two different cartographical projections. (Thompson would demonstrate this principle to children by drawing a normally proportioned dog on a piece of rubber and then stretching it to produce a dachshund.) Again, mathematics showed a path out of the forest of details: Instead of a separate rule to create every difference between two species—a round body versus a torpedo shape, a short snout versus a pointed one, a sloping forehead versus a high brow—you only needed one.

The work was well received—"profoundly interesting," said one biologist—and it may have been the decisive factor in Thompson's election to fellow of the Royal Society of London the following year. This is the highest honor that British scientists can grant one another—not exactly the mark of a crank. It suggests that Thompson's fellow biologists never held him in the low esteem he had assumed. In 1917, Thompson finally escaped Dundee, when the college merged with the nearby university at St. Andrews, and he moved to take the zoology professorship there. He conceded that St. Andrews was, like Dundee, "a cold grey city by the Northern Sea." But the intellectual climate was more congenial. "A town of scholars these five hundred years," it churned out learning like Dundee had once produced cloth. The same year his unified theory of biology, mathematics, and physics was published—he had finished it in 1915, but the First World War had delayed its publication. The book was called *On Growth and Form.* Presenting

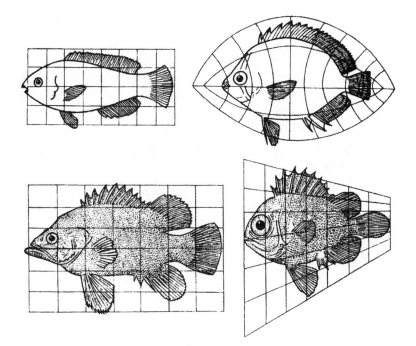

D'Arcy Thompson's transformation grids aimed to show how a simple mathematical operation could turn a parrotfish (top left) into an angelfish (top right), or *Polyprion* (bottom left) into a big-eye (bottom right).
Credit: Reprinted with the permission of Cambridge University.

his philosophy as a tome, rather than a steady trickle of papers, might have been another bad career move, but it secured his place in posterity.

On Growth and Form

If the Portsmouth and Edinburgh papers were sonatas, *On Growth and Form* was a symphony. But all three had the same theme: the path to understanding living things leads through mathematical analysis. Surveying physics and chemistry, it struck Thompson that as these disciplines had become more precise and powerful, they had also become more mathematical. Mathematics, it seemed, was at the heart of good scientific explanation. In this regard, biology lagged far behind the other sciences and was therefore less able to proclaim itself a true Science.

And yet there was plenty of room for mathematics in biology. As he turned his eye to the animal kingdom, Thompson noticed that nature dealt in the same regular forms as the Greek geometers: cylinders, hexagons, triangles. He devoted several chapters to his beloved equiangular spiral, revealing it in the shells of snails and nautilus, in horns, claws, and teeth. He noticed that many of these forms could also be found in inanimate nature, such as waves, hills, clouds, and snowflakes. High-speed photography revealed that a drop of water falling into a puddle made a splash that looked a lot like the polyp of the hydra, a marine invertebrate related to sea anemones. The curly, turbulent eddies that jets of oil or ink make when they flow into water looked like the canopy of a jellyfish. The cells in a dragonfly's wing looked like a film of soap bubbles. Thompson wanted to convince his readers that, in contrast to vitalist philosophy, the laws of physics apply to living organisms and that life does nothing that breaks these laws. What's more, just as physics explained why a wave or snowflake took a particular shape, so too could it explain why an animal did, without recourse to natural selection. Biological form, he reasoned, could be the consequence of physical forces acting on living matter, just as other natural forms could come about through these forces acting on water and rock. In this line of his thinking, Thompson was also pioneering the use of models in biology. He outlined a set of rules and conditions and then saw what the consequences of those rules would be for living matter. These thought experiments produced results that closely matched living structures, convincing Thompson that he understood the underlying principles.

Thompson also pointed out the similarities between natural and artificial structures. An architect designing a bridge will study the loads it must carry, and the gales it will have to withstand, before deciding where to put struts and cables. Likewise the internal structure of a bone is a striking match to the stresses it will endure in life: Its tough outer layer is thicker in the middle than at either end, as fracturing is more likely here, and, like a girder, it contains a network of supports that mimic the forces placed on it. An organism, said Thompson, was "a diagram of forces." Where previous biologists had taken it for granted that organisms were well adapted to their environment, Thompson demonstrated what this meant.

The third strand of Thompson's thinking—and the driving force behind the Theory of Transformations unveiled in Edinburgh and expanded on in *On Growth and Form*—was that mathematical thinking could reveal simple principles underlying the diversity of natural forms. You do not need to assume that a sponge somehow intervenes to determine every aspect of its spicules, any more than a snowflake controls its shape. Both are explained by the rules that describe how minerals aggregate to form crystals. A snail doesn't have a design in mind for its shell; it just adds material to it as it grows, producing the spiral. And small changes in the geometry of this process can produce the many types of shell—thin and fat, simple and intricate, round and spindly—seen in living snails.

Thompson's analyses of living forms, and the comparisons he drew with the design of man-made forms, were visionary. He was, however, stronger on form than on growth, and his advocacy of physical forces as the prime movers in animal development was less successful. For example, he thought that bones develop their networks of struts and supports in response to the physical forces acting on them, rather than by following a genetic program. In fact, a bone will develop its usual reinforcements even if it is transplanted to another part of the body, where it experiences a different set of forces.

His mistrust of evolution led him astray. Bones develop the way they do because natural selection has favored a developmental path that makes bones that can cope with the stresses of life, not because those stresses control the development of bone. His blind spot regarding genetics was at least the equal of any of the earlier comparative anatomists regarding physics. He simply ignored it, dismissing the question of what the units of heredity might be, and how they might be ordered and transmitted, as uninteresting. In his science, as in his career, he could never be bothered to plough anyone else's furrow, and he refused to adapt his thinking. In a 1923 letter he wrote: "The chromosome people are having a good innings; but their theories are top-heavy, and will tumble down of their own weight. It is of little use, meanwhile, to argue with them." Maintaining this position as the century progressed and genetics came to dominate biology must have called on all of Thompson's substantial reserves of bloody-mindedness.

It is difficult to work out what Thompson thought drove growth, but he seems to have believed that the development of animals and plants tended to produce structures that were mathematically harmonious or, at least, that living matter naturally assumed such forms and that elegant mathematics was synonymous with biological adaptation. He replaced the vitalist life force, and Darwin's struggle for existence, with a striving for physical flawlessness: "The perfection of mathematical beauty is such that whatsoever is most beautiful and regular is also found to be most useful and excellent."

D'Arcy was struck, for example, by the ubiquity of spiral shells across the animal kingdom—in molluscs, worms, foraminifera:

> These forms present themselves with but little relation to the character of the creature by which they are produced. . . . We find the same forms, or forms which are mathematically identical, repeating themselves in all periods of the world's geological history; and we see them mixed up, one with another, irrespective of climate or local conditions, in the depths and on the shores of every sea. It is hard indeed (to my mind) to see in such a case as this where Natural Selection necessarily enters in, or to admit that it has had any share whatsoever in the production of these varied conformations.

One might as well invoke natural selection to answer why one type of cloud was common and another rare.

Nowadays, this seems like a failure of imagination. Biologists have no trouble coming up with reasons why natural selection might explain any aspect of anatomy or behavior. But while Thompson's ideas about how evolution works were off the mark, his critiques prefigured many of the issues that exercised evolutionary biologists through the twentieth century. Spiral shells, for example, show convergent evolution, where unrelated animals come up with similar solutions to the same problem—containing a growing body within a shell. The commonness of spiral shells across species and through geological time also hints at evolutionary constraints—the idea that the forms that can evolve are limited by an organism's history and developmental flexibility. Thompson also took issue with the idea that evolution meant progress—a smooth path toward better organisms. If unchanging physical forces controlled animal form, this need not have been the case: "I for one imagine that a pterodactyl flew no less well than an albatross." Now, few think that the giant prehistoric reptiles were stupid, or clumsy, or poorly adapted to their world: Instead we stress the part that chance, in the form of a meteorite, may have played in their demise.

On Growth and Form made Thompson famous. *Nature* compared it to one of Darwin's books, calling it "substantial and stately." *The Observer* newspaper called Thompson "a ripe philosopher and a scholarly historian, possessed of artistic and literary gifts of no mean order." His peers were equally impressed. One Cambridge zoologist wrote to thank D'Arcy: "That the form of animals and plants is not to be regarded as due to a hopelessly complex series of biological factors, but shows the operation of comparatively simple and harmonious physical laws, is, I think, a very great contribution to biology."

Thompson embraced his recognition. In 1918 he gave the Royal Institution's Christmas lectures for children, on marine biology. And throughout his life he seems to have jumped at any chance for a trip that came along—then, as now, one of the perks of academic life. He gave lectures across Europe and in South Africa and the United States, collecting honorary degrees and society medals as he went, and journeyed to Soviet Russia as part of a Royal Society delegation. He also won acclaim as a classical scholar, serving as president of the Classical Association of Great Britain in 1929. In 1937 he represented the Edinburgh Royal Society at King George VI's coronation, and was knighted later that year. He gave radio broadcasts, and produced a stream of pieces on every topic for scholarly journals, newspapers, and magazines. With his love of dancing, taste for Savile Row suits, and habit of walking around St. Andrews with a parrot on his shoulder, it's hard to avoid the conclusion that inside Thompson's lonely soul an inveterate show-off had been itching to get out.

Wartime shortages limited the first printing of *On Growth and Form* to 1,000 copies. Thompson was keen for a new print run but wanted a second edition rather than just a reprinting: "I wish to goodness the first edition would run out, and let me bring out a new one. I have a good deal to correct, and more to add." The Cambridge University Press agreed, writing to him in May 1923, when stocks of the first edition were running low: "[We] will be glad to put in hand a new impression of the work, provided that the corrections you wish to make are few in number." Given D'Arcy's perfectionism and the work's history—the press had contracted for a book of 144 pages; the final work was more than three times this length and arrived several years late—this smacks of hope triumphing over experience.

By October 1929 the press's letters had a plaintive tone: "For the past few years we have announced a new edition as being in preparation, and we have eager inquiries from would-be purchasers from time to time." Thompson delivered the first two chapters of the revised edition in August 1939.

By 1941 his publishers were trying threats: "I must warn you that you have already slightly exceeded the correction allowance specified in Clause (6) of the agreement." The second edition was completed in 1942—it seems to have taken a world war for Thompson to curtail his travels and take to his desk—by which time first editions were changing hands for 10 times the original cover price of 21 shillings. The updated version, in which Thompson struggles to show that he is at least aware of the intervening decades of biology but does not really care to incorporate new developments into his theories, was two volumes and 1,100 pages long—almost twice the length of the first edition. Although well reviewed at the time, it is now thought to be not as good.

After the war ended Thompson, despite being in his eighties, was quick to take to his travels again. In January 1947 he journeyed to India as a representative of the Royal Society. In a typical piece of showmanship, he lectured in Delhi on the mechanical structure of bird skeletons with a live chicken under his arm. But the long journey broke his health, and he had to leave India earlier than he had intended. His health improved a bit when he got home, but the spring of 1947 was unseasonably cold, and Thompson came down with pneumonia. He recovered again in the summer and oversaw publication of *A Glossary of Greek Fishes*, a companion piece to the book on birds published half a century earlier. But pneumonia and other ailments returned in the autumn, and by the winter he was too weak to visit the university. From then on his health gradually declined. Thompson had a sanguine attitude toward old age—unsurprising in one who should be the patron saint of late starters—but as he became an invalid this turned to despair: "I long for release," he wrote on June 9, 1948. He died at home 12 days later. His combined tenure at Dundee and St. Andrews had lasted a record-breaking, if inadvertent, 64 years.

The Last Victorian

D'Arcy Thompson was the last Victorian scientist, in his schooling and classical roots, breadth of knowledge, ferocious work ethic—he would retire to his study at 10:00 each night for two more hours of reading, writing, and thinking before bed—courtly manners, starched collars, voluminous beard, and penchant for a daily cold bath. As Peter Medawar has commented, *On Growth and Form* is a work of natural philosophy rather than modern science. But this was a tactical choice on Thompson's part, as much as a philosophical one. His work on fisheries and oceanography shows that he was perfectly capable of getting his hands dirty in the field and the lab. Yet *On Growth and Form* contains few new observations, and Thompson conducted no significant experiments in this, the enduring thread of his scientific career. When he turned his attention to physics and mathematics in biology, Thompson decided that there were already quite enough facts available—what was needed was a synthesizer to tease the threads of an argument from the disparate work of other researchers. His polymathic background in everything from engineering to Ancient Greek, and his linguistic gifts, which included French and Italian, made him well equipped for the task.

Thompson's unique qualities, unusual career path, and individualistic approach make his science unrepeatable. As a consequence, his place in biology is both marginal and pivotal. His use of mathematics was an inspiring example, but the techniques he favored, based mainly on classical geometry, never took off among his peers. Instead, modern biologists use calculus more often than any other technique. And many biologists remain untroubled by mathematics—today, you can still flick through whole issues of journals in molecular genetics, immunology, and developmental biology without seeing an equation. Thompson's distaste for genetics meant he never engaged with the most successful area of twentieth-century biology—indeed, genetics has filled the role in which Thompson sought to cast physics, by showing that the diversity of living species springs from the same underlying processes. Cell biology and molecular biology also have pursued an understanding of living things by ferreting out the details, whereas Thompson thought such details risked obscuring deeper generalities. He rarely went beyond description to explanation, and the two may have been blurred in his mind. Thompson's favorite way of making a point was through visual analogy; the historian of science Evelyn Fox Keller has pointed out that, for a book about mathematics, *On Growth and Form* contains surprisingly few equations. His transformational grids, the most original part of his work, never took off. It's hard to see how to pursue the method beyond drawing a striking picture, and the analyses required formidable computing power (now that this has arrived, researchers have developed some ways to analyze biological shapes that have a lot in common with Thompson's transformation grids). And some of his ideas were just plain wrong.

All this makes Thompson seem antique, and it's true that as time passes his lines of thought become harder to follow. There are trivial reasons for this: *On Growth and Form* is dense with untranslated quotes in Greek, Latin, and modern European languages. But even with translations, the increasing narrowness of modern scientific training and expertise makes Thompson's all-embracing approach hard for modern readers to interpret. Stephen Jay Gould, who from his introduction to the current edition of *On Growth and Form* seems to have seen himself as a contemporary equivalent of Thompson—and with some justification, as both had wide interests, elegant prose styles, and idiosyncratic views on evolution—wrote that his colleagues saw the book as "an unusable masterpiece."

On the other hand, the debate around vitalism was still alive in Thompson's day, particularly among zoologists—his work helped put an end to it. No serious scientist now believes that living things are exempt from the laws of physics or made from different stuff than dead matter. And although Thompson's mathematics may have gone out of fashion, mathematical models are now used to study everything from enzymes, to bird flight, to the waxing and waning of populations, to where animals should feed, and whether they will evolve unwieldy tails or dazzling, hummingbird-like plumage. Mathematics has allowed biologists to go beyond qualitative verbal arguments to making precise quantitative measurements and predictions and to develop rigorous theories.

Most importantly, Thompson pioneered a new way of thinking about life and a new type of explanation in biology. Underneath the evolutionary eccentricities, *On Growth and Form* challenges biologists to ask why an organism takes the form it does and to look at biology in terms of mechanisms and solutions. Ironically, Thompson's work has helped us understand how natural selection works: Where he talked about an animal being a diagram of forces, biologists now talk about living things being shaped by selection pressures.

And because he made no distinction between the physical and biological worlds, Thompson was also able to pursue this line of thought to what were, in his time, some unusual places. For example, he thought about anatomy at the molecular level and drew links between the shapes of molecules and the shapes of biological structures. Looked at in this light, James Watson and Francis Crick's model of the structure of DNA—the all-time great piece of molecular anatomy—becomes a rather Thompsonian piece of work. Those two were trying to find a biological form consistent with physical forces, an arrangement of atoms and molecules that would be stable. In the process they arrived at a structure that also demonstrates how heredity might work, leaping from form to function. And like Thompson, they arrived at their conclusions by theorizing and modeling, rather than doing experiments—Thompson used cardboard models to try and understand the geometry of cell shapes and bee cells—and they were guided by a belief that the structure of DNA would be aesthetically appealing. As Watson wrote, the double helix was "too pretty not to be true."

But perhaps the main reason that *On Growth and Form* is still in print is that it is simply a beautiful book. Thompson considered writing his principal talent: "The little gift of writing English," he wrote, "is, speaking honestly and seriously, the one thing I am a bit proud and vain of." The spider's web he mentioned in Portsmouth he described as "bespangled with dew, and its threads bestrung with pearls innumerable." Other parts of his writing have a biblical tone:

> Not only the movements of the heavenly host must be determined by observation and elucidated by mathematics, but whatsoever else can be expressed by number and defined by natural law. This is the teaching of Plato and Pythagoras, and the message of Greek wisdom to mankind.

Medawar—himself a superb writer—thought the book "beyond comparison the finest work of literature in all the annals of science that have been recorded in the English tongue."

On Growth and Form should appeal to anyone interested in ideas who appreciates seeing a fine writer making an effort to persuade them. Its status as art has won it a special place in biologists' hearts: They are proud that their discipline has produced such a work of literature, and even its wrongness and eccentricities seem to only add to its charm. In the 1917 edition, for example, Thompson wrote sympathetically about panspermia—the idea that life has traveled between planets and star systems. Propelled by the Aurora, he noted, a microbe could get from Earth to Jupiter in 80 days and reach Alpha Centauri in 3,000 years. It is the sort of book that inspiring teachers press on their students, who then become teachers or researchers themselves and press the book on the next generation of students. From the start, its influence has stretched beyond biologists to engineers and architects, artists, and mathematicians. Soon after the first edition came out, there were inquiries from *The Builder*, the *Journal of Decorative Art*, and the *Mathematical Bulletin*. While writing this chapter in autumn 2004, I visited an exhibition in London of work by the Mexican artist Gabriel Orozco, whose sculpture *Black Kites* features a grid of distorted squares mapped out across a human skull, very much like a Thompsonian transformation. Around the same time, the German architect Frei Otto won the Royal Institute of British Architects' Royal Gold Medal, architecture's most prestigious award. Otto, whose projects include the roof of the Munich stadium used for the 1972 Olympics, pioneered lightweight building using high-tech materials. He took inspiration from cobwebs, bird bones, and crab shells in making his buildings light, economical, and strong, giving them the minimalist elegance of natural forms.

But science is not just about elegance, Aristotle, and the life of the mind. It is also grunt work—prodding, dissecting, and measuring. And while Thompson strained to hear the music of the spheres in the spiral of a seashell, a group of his more experimentally minded contemporaries were attempting to understand living energy by analyzing the chemicals in dog turds, wrapping people in brown paper, and painting stripes on cows.

2   THE SLOW FIRE

D R. COSTARELLI'S INSTRUCTIONS were precise: I should arrive for our appointment by public transport, not the bicycle I usually use for short journeys. I should drink no alcohol beforehand—not such a hardship, as we were scheduled to meet at 9 a.m. More onerously, I should also steer clear of caffeine, so I had to forego the kidney-challenging quantity of tea I pour down my throat each morning. And I should skip breakfast.

So I was feeling sluggish and hungry when she met me and led me to her laboratory, a spare white room in the basement of London's South Bank University. She weighed me, and I lay down on a couch. Her colleague Bill Anderson placed a clear Perspex hood, like a space helmet from a 1950s science fiction film, over my head, and I shut my eyes and relaxed.

For the next 30 minutes, I tried to keep body and mind as inert as possible. I tried—and failed—to remember the last time I had lain still with my eyes closed for half an hour without being asleep. I worried that I wasn't relaxed enough, then tried to halt this train of thought before it spiraled into hyperventilation. I thought of a technique that had once been recommended to me in a massage workshop at the

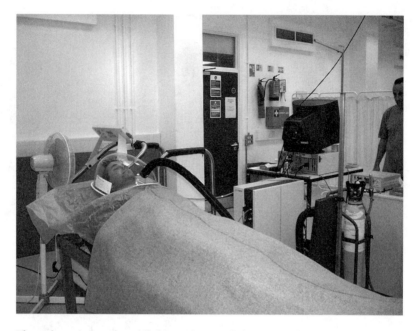

The author tries to relax while his resting metabolic rate is measured.
Credit: Vasiliki Costarelli.

University of California at Berkeley. ("Imagine a golden light moving
up your body, each part of you relaxing as it bathes in the glow.") It was
all very restful—spring sunshine seeped through the blinds, and the
only noises were the hum of the building and a whispered conversation between Anderson and Dr. Costarelli. I tried to stay awake. And
every 30 seconds, tubes in the hood carried some air away to an analysis chamber, which measured the amount of oxygen I breathed in.

For one of those inhaled oxygen molecules, it was a short, simple
trip through my nose and down my windpipe. Then the path forked at
my bronchioles, leading down into my lungs. These air passages divide
again and again, into a labyrinth of ever-narrowing tubes. Eventually,
after crossing more than 20 such junctions, the molecule reached a
blind alley—one of the lungs' air sacs, called an alveolus, where gases
move into and out of the blood. The membrane of an alveolus is no
barrier to a molecule as small as oxygen, and it slipped easily out of the
lungs and into the bloodstream. Instantly, a vastly larger molecule

called hemoglobin, the protein that moves oxygen around the body, snatched up the oxygen. The next part of its journey was chauffeur-driven, riding in a cleft on the protein molecule. Hemoglobin molecules themselves travel around the body in groups of a few hundred, inside red blood cells—oxygenated hemoglobin is what makes blood red. Snug in its new ride, the molecule moved through the bloodstream. Soon it reached the heart, entering the top left chamber and firing out from the bottom left into the aorta, the body's biggest blood vessel. The blood cell began moving through the body in jerks, propelled by each beat of my heart.

Where to next? Any red blood cell could transport its oxygen to anywhere in my body, but demand is greatest in my brain. Even when I am lying on a couch trying not to send any nerve impulses to my muscles or think any stimulating thoughts, this organ consumes more energy than any other—about one-fifth of my body's total requirements, despite accounting for only one-fiftieth of its mass. So say that the oxygen molecule went to my head, traveling up my neck via the carotid artery. Here, it entered a network of blood vessels that get narrower and narrower, like leaving a freeway and turning onto a back street. The blood cell eventually reached a single-track highway barely wider than itself, a blood vessel eight-millionths of a meter across with a wall one cell thick, called a capillary. Capillaries are woven into every tissue in my body. They deliver oxygen and food and pick up waste such as carbon dioxide.

This capillary snakes alongside a nerve cell. Deep in my tissues, there is less oxygen around than there was in my lungs, and in this environment the hemoglobin loosened its grip on its cargo. The liberated molecule was small enough to slip straight through the cell membrane, driven by nothing more complex than diffusion, the tendency for chemicals to move from where they are common to where they are rare.

Inside the cell, the molecule made for a cigar-shaped blob, hanging in space like an airship. This is called a mitochondrion, and it is the unimaginably distant descendent of a bacterium that fused with my unicellular ancestors more than 1 billion years ago. Mitochondria are the cell's power stations. It is there that the chemical reactions of

respiration go on, and they are the reason we need oxygen. The diffusion gradient pulls the oxygen in like a tractor beam. Inside the mitochondrion, after crossing yet another membrane, there are more
molecular machines that break up food molecules, such as glucose,
and use the energy stored in their chemical bonds to fuel life. The main
purpose of these reactions is the production of a molecule called
adenosine triphosphate, or ATP. ATP is the fuel for every cellular
reaction that requires energy. It powers muscles, nerve cells, DNA copy-
ing, and everything else, in animals, plants, fungi, and bacteria. At any
moment, a human body contains only about 10 grams of ATP, but we
make and consume our own body weight of the molecule each day.

The oxygen molecule has come to the mitochondrion to pick up
the trash. It reacts with the electrons and protons that these molecular
machines produce in their fuel-generating work. In the process the
molecule, which consists of two oxygen atoms, breaks apart. Each atom
forms a new alliance with two hydrogen atoms, in a water molecule.
Nothing is ever really destroyed. Each new water molecule will journey
through my veins to my lungs, leave my body as vapor, and float into
the atmosphere. Here it will become part of a cloud, then a raindrop,
and perhaps eventually get taken up by a plant, where it will react with
carbon dioxide to become carbohydrate, creating food for animals.

Aristotle thought that the function of breathing was to cool the
blood. In fact, the opposite is true: By measuring the total amount of
oxygen used by my body, Vasiliki Costarelli is measuring the rate at
which I use energy. This is why I am lying on the couch—to have my
energy consumption, or metabolic rate, measured.

The Torch of Prometheus

To be alive is to be using up and giving out energy. We understand this
instinctively. Many cultures believe an energy field pervades the world
and passes through living beings. Taoists call it *chi*; Hindus, *prana*. To
say that living bodies burn fuel is not a figure of speech. The chemistry
of respiration and that of combustion are identical, and the amount of
energy released from food when you eat it is the same as when you
burn it. Life is a slow fire: Every cell burns fuel to build molecules up

and break them down. Combustion keeps hearts beating, brains thinking, muscles moving, and bodies building.

Chemists came to realize this in the late eighteenth century. The greatest of them all, Antoine Lavoisier, put it best, in 1789, in his *Premier mémoire sur la respiration des animaux* (*First report on animal respiration*):

> This fire stolen from heaven, this torch of Prometheus, does not only represent an ingenious and poetic idea. It is a faithful picture of the operations of nature, at least for animals that breathe: one may therefore say, with the ancients, that the torch of life is lighted at the moment the infant breathes for the first time, and is extinguished only on his death.

Lavoisier studied metabolism throughout his life, until his own torch was snuffed out on the guillotine in 1794, during the Reign of Terror that followed the French Revolution.

A body's metabolism is the work of the tiny furnaces and factories in its cells, which in a human body number billions. These fires can be stoked or quenched. When we exercise, we burn more fuel and get hot. When we eat, our intestines and liver crank up to digest the incoming food, and metabolic rate rises by about a third, which is why Dr. Costarelli told me to come to her lab on an empty stomach. When we get a shock, the pulse of adrenaline released revs up our metabolism to provide the energy for fighting or fleeing. When we wrack our brains over a math problem or a tricky piece of map reading, our neurons demand more fuel. When we are infected, we make things uncomfortable for the microscopic invaders by increasing our metabolic rate and raising our body temperature to feverish levels.

My half-hour relaxing on the couch revealed that, were I to spend all day like that, I would burn 1,726 kilocalories. (The "calorie" counts given on food packaging are, more often than not, kilocalories. The standard scientific unit of energy is now the joule, which equals 0.24 calories.) If I did nothing all day but sit at my computer writing this book, the fidgeting and extra mental activity would probably push that to a little over 2,000 kilocalories. If I blew the day off and went for a hike, or blew this career off and took a job on a building site, that number would rise above 3,000. The kings of energy consumption— *Tour de France* cyclists, wildfire fighters, and polar explorers—need about 7,000 kilocalories a day.

Different parts of bodies demand different quantities of energy. Cells that burn a lot, such as muscle and liver, have more mitochondria. A human liver cell has about 800 mitochondria; the average is about 100. Fat burns less energy than muscle, which is why a woman will usually have a lower metabolic rate than a man of the same weight, because a greater proportion of a woman's body is fat.

Metabolic rate can vary on timescales longer than fright or fever. A cell in a hibernating animal is like a mothballed factory: It keeps ticking over, but only just. Fewer protein molecules are made and broken down, and smaller amounts of substances pass in and out of the cell. Molecular security guards come in to keep key pieces of equipment safe. Energy consumption can drop by 90 percent.

The body can also regulate its weight by tuning energy consumption. Fat cells release a hormone called leptin, which raises energy expenditure, so burning off fat, part of a beautifully precise system which means that, although each human in the developed world eats about 1 tonne of food each year, our body weight over the same period changes by only a tiny fraction of this, if at all. Of course, our biology can only adjust to our diets up to a point. Humans evolved in environments where feeding oneself took strenuous effort and where periodic starvation was common. In affluent Western societies—and an increasing number of developing countries—our instincts and tastes lag behind our abundant food and sedentary lifestyles, leading to today's well-publicized spike in obesity. We combat this trend by becoming acutely aware of our energy budgets, finding out how much fuel our diets contain, and how rapidly different forms of exercise burn it off. Our metabolic rate also declines as we get older—partly because mitochondrial performance drops off—which is why it is harder to stay thin in middle age. The team at South Bank University most commonly measures metabolic rate to help obese people plan their diets. The same measurement is used to ensure that hospital patients who have had major surgery, or been severely burned, get enough food.

In concert with hormones such as leptin and adrenaline, the other main controller of metabolic rate is the nervous system. Mostly the brain does this without troubling our consciousness, but Buddhist monks, through deep meditation, can reduce their metabolic rate to just a third of its usual level. On the other hand, as I scarf down a Danish pastry in the university canteen, Dr. Costarelli tells me that she was once thrown out of a meditation class for passing notes. Restless, high-energy personalities such as hers will consume more energy than more laid-back people.

But the story of my visit to Dr. Costarelli's lab is about more than discovering whether that Danish will go straight to my hips. This morning Dr. Costarelli is aiming to uncover a metabolic rate that underlies the kerfuffle of exercise, shocks and surprises, big meals, and psychoactive chemicals. This is why I laid off the booze and tea, postponed breakfast, and laid down. The South Bank team is trying to measure how much energy my body uses to keep its parts and processes running—to keep me alive—and nothing more. This measure is called resting metabolic rate, and it is the central rate of life. Metabolic rate is the conductor that sets the tempo for the orchestra of biological processes—feeding, growing, breeding, living, and dying. For centuries, scientists have sought to understand what controls metabolic rate and how living things use energy.

Counting Calories

The first person to measure life's energy was Lavoisier. In 1777, in collaboration with the mathematician Pierre Simon Laplace, he put a guinea pig and a block of ice in a sealed chamber. The two Frenchmen measured the rate at which the ice melted and collected the carbon dioxide that the guinea pig exhaled. They found that the quantities were closely matched. The principles of calorimetry, as the technique came to be called, have remained unchanged ever since: We measure either the heat radiated by the body, the chemicals exchanged with the environment, or both. The first calorimeter big enough to carry a human was built in Munich in the 1860s by Carl Voit. The subject sat in a chamber surrounded by water, and Voit measured the rise in water temperature caused by the subject's body heat. Like Lavoisier, Voit was a chemist who became interested in human biology, and his laboratory produced the first generation of biologists interested in metabolism and nutrition.

One of these early biologists, Max Rubner, did more than anyone to lay the foundations for the modern science of metabolism and diet. He was an ardent measurer—in the 1880s he began a daily record of the number of steps he took, with the intention of recording his physical decline with age. He was also a gifted instrument builder, a talent perhaps passed down from his locksmith father, and took calorimetry to new peaks of precision. (Dr. Costarelli's machine is descended from Rubner's calorimeters.) Rubner left Munich at the age of 31, to run his own lab in Marburg. There, in 1889, he conducted probably his greatest experiment. Rubner kept a dog in a calorimeter for 45 days, comparing the food the animal ate with the energy it gave off as body heat and in chemical form via respiration, urine, and feces. The heat measurement came to 17,349 calories, the chemical one to 17,406—near-enough identical. This result showed that animals obeyed the first law of thermodynamics—energy cannot be created or destroyed, only converted from one form to another—and struck a powerful blow against vitalism. Rubner proved that you are what you eat.

Rubner's experimental gifts were matched by the insight needed to turn measurements into theories. In the process he was the first to realize many things that seem obvious to us now. For example, in 1878 he determined that no one foodstuff is the sole source of the body's energy—protein will do as well as fat or sugar—and calculated the energy values of these different food groups (a gram of protein yields 4.1 calories, one of fat 9.3, and one of carbohydrate 4.1). These measurements were used worldwide in analyzing the diets of populations and making nutritional recommendations. It was Rubner who discovered that the amount of energy released by burning food is identical to the amount released by eating it, an experiment replicated in school science classes to this day (at my school we measured the calories in a burning peanut). He found that measuring gas exchange is as good a gauge of energy consumption as measuring heat output, which is why I had to stick only my head inside the machine that measured my metabolism. He showed that chemical reactions can heat bodies in the same way as physical exercise and that animals in cold environments can exploit this principle by raising their metabolic rate to keep warm.

He also believed that his science should be practical, and advocated the public benefits of a scientific approach to eating. Rubner came up with some of the first dietary recommendations: After studying laborers and finding that they burned 3,100 calories a day, he recommended that their daily diet contain 127 grams of protein (later shown to be rather high). During the First World War, by which time he was working in Berlin, Rubner tested different types of flour to see whether, in times of shortage, flour bulked with bran could make nutritious loaves. Whenever he went to a restaurant, he would take a copy of the menu to track changing fashions in diet.

Rubner's experimental labors, keen insights, and public spirit were backed up with a forceful personality. Acquaintances described a well-built man with a powerful physical presence. Photographs show that he was strikingly handsome, with clear, confident eyes and smooth, boyish features under his nineteenth-century dress and beard. But Rubner could also be difficult company, prone to long silences punctuated by bursts of sarcastic wit. He gave no quarter in claiming priority for himself, criticizing others' work, or standing up for his own. A representative sentence from his major book *Die Gesetze des Energieverbrauchs bei der Ernährung* (*The Laws of Energy Conservation in Nutrition*) reads: "A number of arguments, which shall be shown to be perfectly meaningless, have been raised against my methods." (Scientists do not write like this anymore, at least for public consumption, but many will still talk this way with little or no provocation.) He fell out with many colleagues—including Voit, who stymied his work on the energy found in different foods—but was loyal to those he trusted. Rubner's prickliness didn't stand in the way of him having a long and distinguished career, but it may have cost him the biggest scientific prize. His obituary quotes a friend as saying: "You should have won the Nobel Prize, you must have stepped on X's toes." "I did," he replied. (X's identity is not recorded, but may have been Voit.) While in Marburg, Rubner also came up with the first major theory to explain the value of resting metabolic rate.

Max Rubner's Big Idea

Rubner's theory was based on body size. The idea that an animal's size is the best guide to its energy requirements makes intuitive sense. Just as buses need more fuel than cars, and a cottage has a smaller electricity bill than a skyscraper, larger animals are going to have to eat more and so will have a higher metabolic rate. This also means that it becomes harder to lose weight the more weight one loses, as the newly slimline body has a smaller appetite for energy.

But although larger animals have larger absolute metabolic rates, when you consider relative metabolic rates, the opposite is true—in animals, small is not necessarily economical. A shrew must eat more than its own body weight every day just to survive. An elephant makes do on food weighing about 3 percent of its weight, and elephants eat indigestible vegetation, not the high-energy insect diet that shrews need. The smallest bats eat like shrews, and hummingbirds rely on sugary nectar. Some hummingbirds and bats also save energy by dropping into a hibernation-like torpor overnight.

Why, pound for pound, does a canary need 25 times more energy than a cow? Rubner believed that smaller animals' relatively greater need was due to their relatively larger surface area. To understand this, put aside all the other jobs that metabolism does for a moment and consider only its heat-generating powers. Think of an animal as a boiler that burns fuel to keep warm. Heat is generated throughout the body's bulk, in every cell, by the chemical reactions encountered earlier. But it is lost only through those bits of the animal exposed to the outside world. To understand how surface area (skin) changes as a body's size—and hence its mass and volume—increases, imagine a cubic creature, whose sides are each 1 centimeter long. The animal has a volume of $1 \text{ cm} \times 1 \text{ cm} \times 1 \text{ cm} = 1 \text{ cm}^3$ and will lose heat across each of its six sides, which have a combined area of $1 \text{ cm} \times 1 \text{ cm} \times 6 = 6 \text{ cm}^2$. Now imagine the cube grows to double its original size. Its volume will be $2^3 \text{ cm} = 8 \text{ cm}^3$, and its surface area will be $2^2 \text{ cm} \times 6 = 24 \text{ cm}^2$. For an eightfold increase in volume and mass, assuming the big cube is made of the same stuff as the small one, its skin area has increased only four times. Instead of having 6 cm^2 of surface for every cubic centimeter of volume, it has 3.

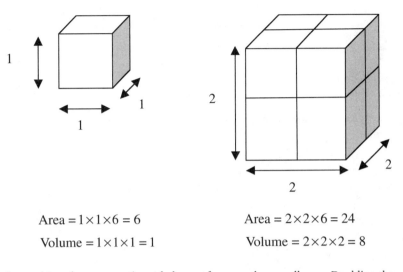

Area = 1×1×6 = 6   Area = 2×2×6 = 24

Volume = 1×1×1 = 1   Volume = 2×2×2 = 8

Large objects have proportionately less surface area than small ones. Doubling the linear dimensions of this cube increases its volume by eight times but its surface area by only four times.

This, thought Rubner, is why a 3-kilogram cat doesn't need 100 times more calories than a 30-gram mouse. The cat has *proportionately* much less skin than a mouse. More of the cat's body is snug and warm away from the outside world, and so it loses heat much more slowly. Nor are mice cooler than cats. The body temperature of mammals and birds, the two groups of animals that burn fuel to keep warm or cool off, does not vary much with size. It follows that small animals must be working harder to heat themselves, and so need proportionately more food.

An animal's size is its defining characteristic. We have such strong impressions of how big animals are that they have entered the language—elephantine, for example, or fleabite. This intuition reflects a profound biological truth. Every aspect of an organism's life depends on its dimensions. The relationship between surface area and volume is one of the keys to understanding the effects of size. Over 400 years ago Galileo noticed that big animals have proportionately thicker leg bones than small ones. The key area here is bone rather than skin. A bone's strength depends on its thickness, which, like surface area, increases with the square of length. So as an animal gets bigger, if the girth of its bones remains in proportion with the rest of its body—which, remember, is gaining weight at the rate of length cubed—the bones will become too weak to hold up the animal. This is why a rhino's legs are short and stumpy and a gazelle's legs are long and slender.

Galileo realized that big animals need more support because gravity exerts a stronger pull on them. "A dog," he wrote, "could probably carry two or three such dogs upon his back; but I believe that a horse could not carry even one of his own size." Or, as the British biologist J. B. S. Haldane put it 350 years later, in his classic essay *On Being the Right Size*: "You can drop a mouse down a thousand-yard mineshaft; and, on arriving at the bottom, it gets a slight shock and walks away, provided that the ground is fairly soft. A rat is killed, a man is broken, a horse splashes."

"Comparative anatomy," Haldane continues, "is largely the story of the struggle to increase surface in proportion to volume." Relative surface area decreases so quickly with increasing size that the bodies of all but the smallest and flattest animals need to make extra surface. (Haldane and D'Arcy Thompson spoke on the biology of size at the 1926 British Association meeting. "Bobbed-haired girl scientists, unable to find chairs, sat on the floor, and additional chairs had to be brought in," said one newspaper report of the event. "Roars of laughter from professors as well as from girl scientists punctuated the discussion.")

We interact with the world across surfaces. We take oxygen across the membrane of our lungs, absorb food across our intestines, secrete waste into our kidneys. The surfaces of all these organs are incredibly convoluted, to create the large surface needed to supply a large body. This is why the journeys through the lungs and bloodstream are so complicated. Both are structured to increase the surface area over which cells can absorb chemicals. A pair of human lungs spread out flat would cover a tennis court. This trick is repeated at every scale of life. Mitochondria have convoluted internal surfaces, to give the maximum area for burning oxygen. Combustion outside cells works in the same way, creating a risk of fire or explosion in any place where there are large quantities of dust, such as coal mines, flour mills, and custard-powder factories.

Of Poles and Penis Bones

In the mid-nineteenth century, several biologists independently drew links between a body's surface area and the rate of its heat loss. The earliest published report dates from 1839; it is by a French multidisciplinary research duo consisting of the mathematician Pierre Sarrus and the biologist Jean-Francois Rameaux. Sarrus and Rameaux were also interested in energy requirements: They were calculating the food needs of employees at the state tobacco factory in their native Strasbourg.

Eight years later the German biologist Carl Bergmann realized that the link between size and energy could control how an animal fit its environment, as well as its internal economy. He further deduced that the physical environment could produce patterns in the living world, thus showing that an understanding of how individual organisms work can lead to an understanding of how nature as a whole works.

In 1847, Bergmann suggested that, in any group of closely related, warm-blooded animals (also known as endotherms, because they heat themselves from within, or homeotherms, because they maintain a constant temperature), species living in cold climates would be bigger than their relatives from hot places, to help them conserve heat. So animals living near the poles will be bigger than those at the equator. A brief mental tour of the animal world provides plenty of examples to support this notion. Arctic foxes are the smallest animal living on the Arctic island of Spitzbergen; they are considerably larger than the red foxes that live in my London garden. Polar bears in the Canadian arctic are bigger than Californian black bears. Emperor penguins in Antarctica weigh about 30 kilograms, but equatorial Galapagos penguins weigh on average less than 2.5 kilograms. There is support for the rule within species, too—the wolves of Alaska are bigger than those of Arizona. Humans too are bigger away from the equator. Conversely, large animals in hot climates can have a problem losing heat, and so have evolved body structures that help them cool off, such as the high-surface-area radiators attached to either side of an elephant's head. In the 1870s the American biologist Joel Allen extended Bergmann's arguments to animal shape, suggesting that animals living in hot climates should be longer and thinner, with larger appendages and extremities. Desert jack rabbits have long legs and ears and slender bodies; Arctic hares are squat.

Not every appendage shrinks toward the poles. Many mammals have a penis bone, which augments an erection—a feat male humans must accomplish through blood pressure alone. Two Canadian zoologists, Steven Ferguson and Serge Lariviere, measured the penis bones of 122 species of carnivorous mammal and found that the more polar the beast, the bigger the penis bone, relative to body size. Elephant seals living in temperate waters weigh a couple of tonnes and have a penis bone about 30 centimeters long. Male walruses swimming in the Arctic Ocean, on the other hand, weigh in at a comparatively flyweight 1.7 tonnes, but have a penis bone a whopping 60 centimeters long.

It's not swimming in frigid polar waters that leaves the walrus needing a little something extra, the Canadian duo believe. Populations of polar species tend to be more thinly spread than those in more hospitable climates. Elephant seals live in big colonies and use their extreme bulk to win and defend mates. For a male walrus, maintaining such a harem would be a geographical impossibility—so, when he does meet a female, there's a heavy pressure to perform. Ferguson and Lariviere believe that a long penis bone helps males who mate sporadically perform well in the competition between males to sire offspring.

The way that a species changes size through time also seems to show the workings of Bergmann's rule. A study of the changing sizes of American woodrats over 25,000 years (based on the size of their droppings) showed that the animals got bigger when the climate was cold and smaller when it got warmer. Other studies, using museum specimens collected over the past century, have suggested that several bird species have gotten smaller as the world has warmed—showing that the effects of global warming will be subtle and surprising, as well as potentially Earth-changing.

So the logic is simple. Tropical and desert animals should be small and lanky; polar ones should be big and rotund. But Bergmann's rule is controversial. Ecologists still can't agree whether it holds or not, despite testing it on everything from robins to moths to kangaroo rats. Some have asserted that the rule fails on the grounds of evidence, and that the majority of animals don't show a tendency to get bigger toward the poles. One 1936 study claiming that the geographical trends in the sizes of American mammals showed support for the rule was later attacked for taking the dubious shortcut of using numbers from field guides, rather than new measurements. Others have taken up conceptual cudgels, arguing that large size is not actually a good way to conserve heat, compared with, say, thicker fur. Some contend that big animals may actually have a harder time keeping warm in cold climates because they need more food overall than small ones. Another camp believes that the patterns in body size noted by Bergmann are genuine but that they are not caused by temperature. Humidity is more important, some claim. Others have asserted that big animals are better placed to cope with a seasonal climate because they can carry larger fat reserves and so survive periodic brushes with starvation. This might explain why some species of reptiles and insects, animals that don't need to maintain a constant body temperature, also follow Bergmann's rule.

The past century has seen regular to-ing and fro-ing in scientific journals, as researchers have taken potshots at each other's hypotheses. Recently it has looked as if the debate might be going Bergmann's way. A 2003 review of the evidence, by Shai Meiri and Tamar Dayan, from studies covering a total of 94 bird and 149 mammal species found that about three-quarters of the birds and about two-thirds of the mammals obey Bergmann's rule. It looks as if warm-blooded animals *are* bigger nearer the poles, although why that is so is still open to debate and is probably the consequence of many different evolutionary forces working at once.

Two-thirds might seem like not terribly impressive support. The law of gravity would read rather differently if apples fell up, or sideways, one time in three. But biological rules such as Bergmann's, of which we shall be meeting many, are not the same as physical laws such as gravity, which hold true everywhere and which can be used to make precise predictions. Biological rules are more often trends that, all other things being equal, hold more often than not. One or several exceptions are rarely enough to discredit the whole pattern. For example, the Galapagos penguin lives closer to the equator than any other penguin, but it is only the second smallest of its kind. The smallest, the fairy penguin, lives in Australia. Temperature is not the only thing that affects body size; animals have got to do lots of other things besides keep warm. There's no point being huge if you want to fly, or swing from one tree branch to another, or escape from your predators down a burrow. Meiri and Dayan found that Bergmann's rule holds less well for migratory birds than sedentary species, perhaps because these species avoid cold climes by traveling. And, they found, rodents are more likely to buck the rule than other mammals such as bats, perhaps because rodents are more likely to live in insulated burrows.

This untidiness makes it easy for researchers to disagree about patterns in nature such as that spotted by Carl Bergmann. The gap between the trend and the variation becomes disputed territory between the lumpers who see patterns and the splitters who see diversity. In 1956 the great German evolutionary biologist Ernst Mayr said that patterns such as Bergmann's rule should be taken as valid if they applied to more than half of all species. This rather unambitious yardstick brings to mind the scientists' joke on the differing standards of proof demanded by different disciplines. An astronomer, a physicist, and a mathematician are taking a train ride through the Scottish highlands. From the train window the astronomer—being of an observational bent but also prone to sweeping generalizations—spots a black cow standing in a field. "Look," he points out, "cows in Scotland are black." The physicist corrects the astronomer: "You can't assume that," she argues. "All we can really say is that particular cow is black." The mathematician, believing only what can be proved beyond any doubt, rolls his eyes: "Really," he sighs. "All we know is that one side of that cow is black." If this party of caricatures had included an ecologist, he would probably have asserted that every field in Europe contained a solitary black cow. This isn't to say that the patterns in nature aren't real or that the explanations for them are unscientific, just that we can't expect things to be too neat.

The Surface Rule

So an animal doesn't burn energy twice as fast as one half its size or three times faster than one a third of its size. At what rate does relative metabolism trail off, and why? One obvious idea is that the decline in relative metabolic rate mirrors the decline in surface area relative to volume. If this were the case, every 1,000-fold increase in body mass would see a 100-fold increase in metabolic rate. In 1883, Rubner, based on measurements of the metabolic rates of seven different-sized dogs, concluded that this was indeed so. He calculated the surface area of his experimental hounds using a formula published four years earlier by another German, Karl Meeh. Meeh encased human experimental subjects in paper cylinders—I imagine them looking like the Tin Man in *The Wizard of Oz*—and then weighed this paper. Knowing the paper's thickness, he could work out the area of the people. Meeh concluded that to work out an animal's surface area you should first weigh it, then take the two-thirds power of this number, and multiply it by a constant (that is, surface area = constant × weight$^{2/3}$).

Meeh's experimental conclusion matched the predictions of basic geometry. Let's return to the cubic beast. Its volume—assuming, again, that body mass reflects body volume and that the flesh of different animals is equally dense—is proportional to its length cubed. This is where heat is generated. Its surface area, where heat is lost, is proportional to its length squared. To get from 3 to 2, from a volume to an area, we need to multiply by 2/3 ($3 × 2/3 = 2$). This is the relation between volume and surface area—so Meeh's experiments produced the same result expected from mathematical first principles. (Raising mass to the power of two-thirds is the same as squaring it and then taking the cube root of that result. For example, $10^{2/3} = \sqrt[3]{10^2} = \sqrt[3]{100} = 4.64$.)

The formula also introduces the important concept of similarity. Similar shapes or objects are those with identical proportions, such as the two cubes with sides 1 and 2 centimeters long. To get from one to the other, you multiply every dimension by the same amount. All cubes are similar. If you multiplied the different dimensions of a cube by different amounts, you would get a different solid, such as a cuboid or a trapezoid. A formula that calculates area as the two-thirds power of mass assumes that large animals have the same proportions as small ones, that a cat is a scaled-up version of a mouse. This is obviously false: The proportions of animals change as they get bigger, both within and between species. But the assumption of geometric similarity might be accurate enough not to upset experimental results over the relatively small size range shown by Meeh's humans or Rubner's dogs.

When Rubner applied Meeh's formula to his dogs, his measurements showed that the animals' energy output was proportional to—similar to—their surface area. The measurements made good theoretical sense. Rubner collected evidence on his proposed surface law by experimenting on animals ranging in size from a horse to a mouse. On retiring in 1924 he wrote that he considered the law his greatest contribution to science. "There can be no doubt about [its] general validity," he concluded.

Rubner's surface law is summed up thus: *Each square meter of the surface of every mammal produces the same amount of heat.* He estimated this heat output at about 1,000 calories per square meter of skin per day. It was in this sense, Rubner believed, that large and small animals were similar. Although the idea is most closely associated with Rubner, the French biologist Charles Richet came to the same conclusion around the same time, based on measurements of rabbits. This is how science proceeds, usually, not by lone geniuses striving toward stunning breakthroughs but by a community inching forward across a broad front. (Richet, incidentally, was probably happy to let Rubner have the credit for the surface law—Richet did win a Nobel Prize, for his discovery of anaphylactic shock, which laid the foundation for our understanding of allergies.)

The surface area law soon became widely accepted by biologists. Scientists wishing to understand metabolism at the beginning of the twentieth century were thus faced with measuring two variables: metabolism and body surface area, the former by either heat output or respiratory exchange, the latter by either direct measurement or some approximation formula. Physiologists went about the task with zeal, measuring the surface area of thousands of men, women, children, babies, fat people, thin people, Asians, Westerners, the able bodied, and the handicapped—one study included two people who had each lost both legs in train accidents and "a 36-year-old cretin," cretinism being the result of an underactive thyroid gland, which also leads to slow metabolism. Measurements of metabolic rates were made for those

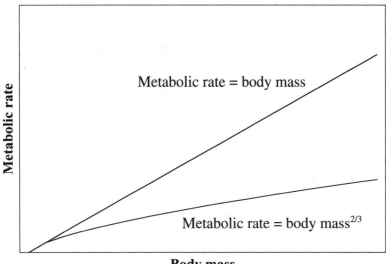

The surface law hypothesized that every square meter of mammal skin produced the same amount of heat. So relative metabolic rate would decline at the same rate as surface area relative to body volume.

lying down, sitting down, moving around, diabetics, hyperthyroids, and so on.

As this project progressed, researchers taking measurements of skin surface found that Meeh's formula seemed to overestimate the body's area. Meeh had evidently demurred from wrapping his subjects too tightly. Later investigators showed no such squeamishness. They stripped their subjects down to their underwear and wrapped them in tin foil or plaster of Paris. Others took photographs from every angle or worked out the skin's area from its electrical conductance.

Despite its inaccuracies, Meeh's formula was in common use for nearly 40 years—probably because it was mathematically simple and required only one measurement, weight. It eventually crumbled under an assault by Eugene DuBois, a New York–based pathologist, around 1915. At least as far as humans were concerned, DuBois was perhaps the foremost surveyor of skin and measurer of metabolism.

Basal metabolic rate is a slippery concept. It's hard to define what the basal functions of life are and impossible to measure them in a living, breathing, fidgeting animal without also measuring the energy used up by lots of other processes and activities. Some researchers now avoid the term basal metabolic rate altogether, while others use it to refer to sleeping energy consumption; the measurement Dr. Costarelli made on me is usually called resting metabolic rate. DuBois helped develop the standards for measuring metabolic rate—the subject should be relaxed and hungry, at a temperature where he or she is neither shivering nor sweating—in an attempt to find some sort of baseline for metabolism. The hope is that the imprecisions even out over many measurements. Indeed, this is a general issue in biology. In a reversal of the earlier joke, a physicist can know the mass of every electron in the universe by weighing one. A biologist weighing a dog knows only the weight of that dog on that day. This variability has made the outlook of some biologists more similar to the mathematician than the astronomer in that joke, stressing the uncertainty in nature, rather than the generality.

If scientists could standardize their measurements and comparisons of energy consumption, they could make progress in understanding metabolism. Minimizing metabolic rate before its measurement was an obvious option—it would avoid all the problems of defining what might be "typical" metabolism. But searching for one catch-all measurement casts the validity of the enterprise into doubt. No human, or any other animal, spends all its time in repose. As we have seen, metabolic rate varies hugely depending on where you are and what you are doing. So what can lying down for half an hour really tell us about energy consumption throughout life? Does a standardized laboratory measurement really tell us anything about the real world? Fortunately, it looks like it does. Metabolic rates of active animals measured in the field are, more or less, three times higher than their resting metabolic rate. A wild animal's maximum metabolic rate is usually about 10 times its resting rate, although highly trained human athletes and racehorses can achieve up to 20 times their resting rate. It seems that comparisons based on resting rates can be useful guides to energy needs in the wild.

Eugene DuBois, working with his engineer brother Delafield, tried to nail down human surface area once and for all. Eugene was a man of action as well as science. He served in the U.S. Navy Medical Corps in both world wars and was awarded the Navy Cross in the first, for protecting the crew of the submarine he was serving on from chlorine poisoning. DuBois was interested in how the body performed at extremes. He was a leading light in aviation and diving medicine, and in 1928 he conducted an experiment on the arctic explorer Vilhjalmur Stefansson that showed it was possible to live for several weeks on meat and water alone with no apparent ill effects—as the Inuit do and as Stefansson had done on some of his journeys—an experiment often cited by contemporary advocates of low-carbohydrate, high-protein diets. DuBois, who had a penetrating glare and hawklike features, looks like he too might have liked to live on meat and water alone. He does not look like a man who would have let an experimental subject's dignity get in the way of an accurate measurement.

The DuBois brothers began by wrapping their subjects in brown paper, making molds that could then be laid flat on photographic paper. When exposed, the areas covered by the mold could be cut out and weighed, and this weight could be converted into area. In 1915, they published a formula for calculating body area from 19 different measurements—everything from circumference of the knee joint to length of the hand. Medics ignored this method, probably because it took about 10 minutes to do all the measurements and calculations. The DuBois brothers went away to measure more people's surface area. By now their quest for accuracy had led them to replace the paper wrapping with sticking plaster, and to make their molds by tipping paraffin wax over the mummified victim. Eugene and Delafield returned a year later with a simplified formula that called only for height and weight to be measured.

Other researchers developed less sweaty ways to measure surface area—after all, unlike humans, animals can't be expected to submit meekly to being transformed into a waxwork. A team at the University of Missouri, led by Samuel Brody, came up with a device like a paint roller that they called the surface integrator. Once the animal was covered in stripes, the number of revolutions multiplied by the circumference of the roller gave the body area. In 1926, Brody's team used the integrator to measure the surfaces of 600 cattle. In 1927, Hannah

Samuel Brody's surface integrator being used to measure the area of a cow.
Credit: Brody Environmental Center, University of Missouri.

Stillman Bradfield, a graduate student at the University of Missouri, published a study describing her use of the surface integrator to measure the areas of 47 naked young women—to see how their areas compared with the men from whom most of the formulas had been derived. (Eugene DuBois was unimpressed: "There still remains some doubt as to whether the integrator is as accurate as the method employing molds," he sniffed.)

The surface law was going strong. Succeeding generations of physicians have tinkered with the numbers in the DuBois formula, but calculations of surface area are still based on height and weight. Many medical quantities, such as drug dosage, intravenous feeding rates, calorie requirements, and heart output are still calculated in terms of body surface area, as calculated by these formulas. As we shall see, there is no good reason for this.

3   MOVING THE LINE

EVEN BY THE WEIGHTY standards of nineteenth-century scientific German, Max Rubner's *Die Gesetze des Energieverbrauchs bei der Ernährung* was a notoriously indigestible slab of prose. But Max Kleiber, who was sent a copy by his professor, was one of the few people to read it all the way through. Kleiber had nothing better to do at the time—he was in prison.

To call Max Kleiber a free spirit hardly does him justice. Serving in the Swiss army during the First World War—Switzerland was neutral but mobilized its military to defend its borders—he was dismayed to learn that senior Swiss officers had been passing information to the Germans. Believing that the chain of command had been corrupted, he felt he could no longer pass on orders; Kleiber ignored his next call-up and was arrested and jailed.

The initial military summons had found him in Canada. After a year of university in Switzerland, during which he scandalized Zurich society by walking about town hatless, in sandals, and with an open collar, Kleiber decided the academic life was not for him and that he wanted to get as far as he could from people and their works. Along with two friends, he tried to make a life as a homesteader in Alberta,

Max Kleiber (1893–1976).
Credit: University of California, Davis.

Canada, the most remote and least populated place they could afford
to get to. When he arrived, however, he found that he had over-
estimated his disdain for society. "This experiment made me aware of
the degree to which *Homo sapiens* is a social animal," he wrote, and
tutoring his neighbor's daughter in mathematics turned out to be more
rewarding than eking out a living from the land. However, Kleiber was
reluctant to give up the agrarian ideal and had another go at farming

back in Switzerland, once his four-month prison sentence was over. He founded a commune with a group of like-minded conscientious objectors. But arguments within the radical collective, and his own realization that he was a better scientist than farmer, sent him back to university. His prison term had led to his expulsion from Zurich's College of Agriculture, but the same professor who had sent him Rubner's book pulled strings to get Kleiber readmitted.

In 1929, Kleiber emigrated with his wife and daughter to the United States, to work at the University of California's agricultural college in Davis, a town about two hours' drive inland from San Francisco. He got the Davis job only because a more eminent European animal physiologist wanted more money than the University of California was willing to pay, but he ended up spending the rest of his life there. Neither emigration nor age dimmed his radicalism. In 1949 he became the first University of California professor to refuse to sign an anticommunist loyalty oath, which the university heads had demanded from all their employees. He demonstrated against the atom bomb and took on Edward Teller, the father of the U.S. nuclear weapons program, in public debate. Late in life he protested against the Vietnam War. "I seem to have a greater than average allergy against mental schisms," he wrote. "This means an enhanced tendency to express my ideas through actions." He now has a building on the Davis campus named after him.

Davis originally hired Kleiber to build and use equipment to measure respiration and metabolic rate in cattle. This is an important agricultural question because understanding how metabolism changes with size helps farmers calculate how much feed their livestock need and the rate of growth and the quantity of milk and meat that they should expect from an animal. Soon after he arrived at Davis, Kleiber used these measurements to provide biology with one of its few universals and one of its most enduring mysteries.

Cracks in the Surface Law

In the late 1920s, Rubner's surface law of metabolism was dogma. Scientific studies often gave metabolic rate as a function of body surface area alone, without reporting body size or height. Studies that didn't conform to the law's predictions were usually dismissed as inaccurate. But the anomalous measurements were beginning to stack up, and researchers were realizing that previous studies of both metabolism and surface area might not be as accurate as they had thought. For a start the techniques and procedures used for measuring metabolic rate had moved on since Rubner's day. Rubner's calorimeters were excellent, but his experimental set ups were not the best suited to getting consistent, reproducible results. Rubner's sample sizes—one experiment involved 2 men, 5 dogs, 5 rabbits, 3 guinea pigs, and 12 mice—would today be seen as too small to allow for firm conclusions. During his time, beer was a routine part of experimental diets—now a strict no-no, as I discovered when I had my own metabolic rate measured. His favored period for taking measurements was 24 hours, a long time to expect a dog to sit still. And his favored experimental temperature was 16°C; his animals must have been shivering to keep warm.

Eugene DuBois said this about the experiments underpinning the surface law: "Like much of Rubner's work, the experiments and measurements contained many mistakes, but Rubner was a genius who drew correct conclusions from data inadequate for any other man." DuBois was reluctant to let the surface law go, which was not surprising given all his work on surface area, but even his backhanded compliment to Rubner's methods came to look increasingly optimistic.

This isn't to say that Rubner was a fraud or an incompetent. Like many a scientist before and since, he may have been reluctant to let the data get in the way of a good theory, but he and the other biologists taking these measurements in the nineteenth century were exploring uncharted territory. You can now buy off-the-shelf apparatus and software to measure metabolic rate and process the results, but Rubner's generation had to find out for themselves what worked. There were no standard recipes to follow. For example, it became common to measure metabolic rate in fasting animals. But how long should this fasting period prior to measurement be? A rat will empty its stomach more quickly following a meal than a dog will. A ruminant such as a cow never really stops digesting, and its gut contents make up a good percentage of its weight. The diversity of methods made it hard to compare results coming out of different laboratories.

So it took a while to work out a standard measurement for metabolic rate. But these difficulties were trivial compared with measuring surface area. Each measuring device, technique, and approximation formula gave wildly different results. Skin has many folds and is immensely stretchy—how taut should it be when measured? If you are measuring the skin taken from an animal, should you peg it out for measurement or let it lie slack? Does surface area include internal surfaces that interact with the outside world, such as the lungs and intestines? If so, how do you measure or estimate their area? The exposed area of a living animal is changing all the time, as it curls up for warmth or stretches out. Some biologists took a back-to-front approach, using the surface area measurement that gave the best fit with the surface law. Rabbits' metabolic rates, for example, gave a better fit if you ignored their ears, so they were dismissed as anomalous appendages. DuBois suggested that surface area measurements were reliable as long as the same method was used for each study. But even this turned out to be untrue: Samuel Brody found that, in the hands of three different investigators, the same technique for measuring the surface area of a rat gave answers varying by 60 percent. It was starting to look as if Rubner's conclusions were no more accurate than his measurements.

It was chaotic—Kleiber compared the situation to the measurement standards of the middle ages, when the length of a foot varied from town to town. Biologists began to despair of measuring surface area accurately enough to tell whether it matched an animal's metabolic rate. But when Rubner's law began to fall from favor, it was replaced not by chaos but by a different form of order that was much harder to explain.

From 2/3 to 3/4

Never one to let a received opinion go unchallenged and never afraid to dissent, Kleiber was the first to spot this new form of order. He compiled measurements of metabolic rate in different-sized animals,

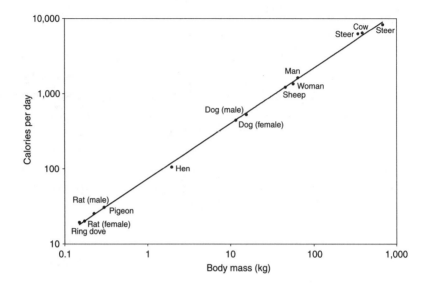

Max Kleiber's 1932 analysis showed that metabolic rate is proportional to body mass raised to the power of 3/4.

ranging from a rat to a steer, and published the results in the Davis house journal *Hilgardia* in 1932, the same year that Rubner died in Berlin. Kleiber's graph from this paper, of metabolic rate plotted against body mass, remains well known among biologists to this day.

Metabolic rate does not correspond simply with body mass, so the graph should be a curve. Kleiber made the curve a straight line by plotting the logarithm of mass against the logarithm of metabolic rate. The notches on a logarithmic axis mark proportions—instead of ticking along from 1 to 2 to 3, they go from 1 to 10 to 100 to 1,000. Because metabolic rate changes at a constant rate as body mass increases, plotting the axes in this fashion produces a straight line, making it a lot easier to compare animals at different points on the graph. The logarithms help us see past the absolute weight difference between two animals to see how many times bigger or smaller one animal's mass and metabolic rate are than another's.

As we can see, Kleiber's comparison revealed a regular relationship between the logarithms of mass and metabolic rate, getting rid of any need for complicated and almost certainly unreliable measurements of body surface area. Although the points for each species on the graph of mass versus metabolism are averages—any one animal is unlikely to lie right on the line—the correlation is nevertheless tight. Correcting for body mass accounts for more than 90 percent of the variation in metabolic rate.

But in cutting through the empirical muddle, Kleiber created a theoretical problem. The gradient of this line is 0.74—nearly three-quarters, not the two-thirds, or 0.67, the surface law predicts. In other words, bigger animals produce more heat per unit of surface area—mice are a bit cooler than cats after all—and bigger animals need more food than the surface law would predict. The surface law predicted that metabolic rate rises 100-fold for every 1,000-fold increase in body mass; Kleiber found a metabolic rise of 180 times for the same weight gain. "Modern American animals take the surface law less seriously than did the earlier European animals," he concluded. "The gadgeteers who designed apparatus for surface measurements, the statisticians who derived formulas for calculating 'true' surface areas, and the theoretically inclined biologists who discussed the proper interpretations of the surface law seem not to have been interested in the reliability of the surface law itself."

To work out how many calories a warm-blooded animal burns each day, said Kleiber, calculate the 3/4 power of its mass (i.e., cube it and then take the fourth root of that number. For example: $10^{3/4} = \sqrt[4]{1,000} = 5.6$) and then multiply by 70. Doing this sum for me, with my weight of 76 kilograms, gives a figure of 1,802 calories per day—very close to Dr. Costarelli's measurement. Kleiber chose three-quarters because the relationship he found of 0.74 is statistically indistinguishable from 0.75, and raising a number to the power of 3/4 was much easier to calculate using slide rules, which at that time were biologists' most powerful computational aid. To translate Kleiber's law into the language of mathematics, metabolic rate = constant × mass$^{3/4}$.

Kleiber might have been the first to publish this relationship in a scientific journal, but as he was later delighted to learn, it was first aired in 1699. In *Gulliver's Travels* the Lilliputian king allocates Gulliver a daily ration equivalent to the diet of 1,724 of his subjects. Assuming that the average Lilliputian was about the same size as Gulliver's index finger, Kleiber calculated that Gulliver was 26 times higher than a Lilliputian and so weighed 26^3, or 17,600 times more than him or her; 1,724 is 17,600 to the power of 0.76.

Allometry

The relationship between metabolic rate and body weight is an example of a biological pattern called allometry, which compares how the value of any biological trait, such as metabolic rate or leg length, changes with the total size of a plant or animal. It shows whether, as things get bigger, they become proportionately bigger or smaller—in other words, if their shape changes. For example, baby humans have big heads relative to their bodies. But as they grow up, their heads grow more slowly than their bodies, bringing about adult proportions. Other features get proportionately bigger as an animal's size increases. Deer antlers show this trend: The biggest British species, the red deer, is about 4 feet tall at the shoulder, and its antlers span about 3 feet. The largest deer that ever lived, the extinct Irish Elk, was less than twice as tall, standing at about 7 feet. But its antlers were 12 feet across. In other words, big deer have *really* big antlers. But although the Irish Elk's antlers might seem freakish, the allometric comparison across all deer species shows that they are as big as we should expect for a deer of that size. The power of allometry is that it allows us, literally, to put things in proportion. The examples of babies and deer show that allometric comparisons can be made within species, between species, at different times during an individual's growth, or across a population of animals of the same age. Allometry can be applied to fossils, to study a species' changes in shape through time. It can also show when something is anomalously small or large. Perhaps no other theory in biology can match its sheer usefulness: The data for allometry studies are usually easy to collect and interpret; the analysis is transparent—biologists are not generally good at mathematics, but anyone can understand the allometric equation—and its results can be presented in a visual fashion. Allometry is one of the most powerful techniques for revealing patterns in biology.

To pop back into math-speak: The general form of the allometry equation is $y = ax^b$. If b is greater than 1, the trait gets proportionately larger, like deer antlers; if b is less than 1, as is the case with metabolic rate or babies' heads, it gets proportionately smaller. A good analogy is two bank accounts with different rates of compound interest. Over time the amount of money in each account will diverge, but each account's growth rate will remain the same. If b is zero, the y variable remains constant. If b is less than zero, the trait declines with increasing size. This is what happens with relative, or cellular, metabolic rate. If metabolic rate is proportional to body mass raised to the power of 3/4, then relative metabolic rate will be proportional to this figure divided by simple mass (mass to the power of 1, in other words)—which equals body mass to the power of −1/4. Kleiber helped show that the idea of allometry could extend beyond solid anatomical features such as horns and into the body's processes, such as metabolism. Biological growth, like financial growth, is a compound process—another reason logarithmic axes are more useful than arithmetic ones when comparing different-sized animals. Cells arise from cells, so the amount of new material added in any time period depends on how much there was before.

Allometries are an example of something called a power law, so called because the y variable depends on the x variable being raised to some power. Power laws spread far beyond biology. The frequencies of earthquakes and landslides of different sizes follow them, as do the sizes of air bubbles in a breaking ocean wave or the length of the waking periods during a night's sleep. They also apply to social patterns, such as the length of time patients in the United Kingdom wait for a hospital operation, or the size of changes in financial markets, the frequency with which words appear in a language, and the popularity of people's names. Power laws also relate the size of an event to its frequency. Crudely put, in systems that follow negative power laws, the probability of events such as quakes or stock market crashes declines at a constant rate as those events get bigger. Small earth tremors are common; city-destroying catastrophes are rare. As for sleep patterns, most of the time you will wake up long enough only to turn over and go back to sleep, but occasionally you'll lie awake for what seems like half the night.

Allometry is most closely associated with the British biologist Julian Huxley. Huxley was an intellectual aristocrat. He was the grandson of Thomas Henry Huxley, the Victorian biologist whose robust defense of evolutionary ideas earned him the nickname Darwin's bulldog; the brother of author and philosopher Aldous, whose books include *Brave New World*; and the half-brother of Andrew, who won a Nobel Prize in 1963 for his studies of how nerves work. Julian made the rest of the clan look like slackers. He did important work on animal behavior, genetics, evolution, and developmental biology. He was a busy media scientist, producing a stream of journalism, books (including a volume of poetry), and radio and television broadcasts, and codirecting a pioneering nature film about seabirds, *The Private Life of the Gannets*, which in 1937 made him the only biologist so far to win an Oscar. Huxley was active in social, political, and philosophical debates and campaigned against the bogus reasoning of Nazi race science and Soviet genetics. He argued for better diets and against air pollution. He lobbied for, and helped establish, the first government-supported nature reserves and national parks in Britain. He advocated a humanist view of religion that excluded the supernatural, which caused him to be branded "Europe's leading atheist" in the United States. He ran the London Zoo before and during the Second World War—once chasing down an escaped zebra during an air raid—and became the first director general of UNESCO, the United Nations Educational, Scientific, and Cultural Organization.

In 1924, Huxley published what became a seminal paper showing that male fiddler crabs develop one disproportionately large claw because this claw grows about six times more quickly than the rest of their bodies. He offered the allometry equation as the mathematical description of this pattern. Huxley went on to show that this equation applied to other features of animals' anatomy, including deer antlers, and that different animal forms were the result of different growth rates. By the late 1920s and 1930s, measuring patterns of this sort was one of the hottest areas of biology. Huxley wrote a book on the subject, *Problems of Relative Growth*, and dedicated it to D'Arcy Thompson, the most passionate advocate of the links between mathematics and animal form. Thompson was ungrateful. His letters to Huxley show that he thought he had anticipated this work in *On Growth and Form*—"I never thought it necessary to coin a phrase for it," he wrote—and he took issue with Huxley's use of logarithms and power laws, arguing that simple arithmetic would have been just as good at describing growth. Most biologists, however, followed Huxley's lead.

Noah's Scales

Across the Atlantic, Kleiber was not challenging the surface law single-handedly. Brody too had lost faith, coming to the conclusion that surface area was impossible to measure accurately and bore no relationship to metabolism even if it was. In Missouri his team measured the metabolic rate of more and more species of a broadening range of sizes. In 1934 they published a graph that stretched from a mouse to an elephant. The gradient of this was 0.73, statistically indistinguishable from Kleiber's result. Both, however, were significantly different from the value of 0.67 predicted by the surface law.

By the late 1930s even DuBois was wavering: "The metabolism per square meter shows a pronounced tendency to be increased with the increasing size of the animal," he wrote. "The puzzling question [is] why small animals, like mice, have a lower metabolism per unit of surface [area] than large animals, like horses." As for the answer to this puzzle, DuBois was showing signs of fatalism. "I do not know where or when the various species of animals were given their basal metabolism," he wrote. "Perhaps Noah did it when they left the Ark. I suppose that the reason he could not do a uniform job with the animals was because he did not have scales small enough for the dwarf mice and large enough for the bull and the elephant."

Yet surface area approximations, including the DuBois brothers' formula and updated versions of it, are still used in medicine today, although weight is used more often. Calculations of body surface area are used to prescribe drug doses in chemotherapy treatments. Patients' fluid and energy requirements are often calculated the same way. And some of the body's workings are still most commonly expressed in terms of its surface area. These include the rate at which the kidneys filter waste from body fluids, and the cardiac index, a measure of how hard the heart is working, which is defined as liters of blood pumped per minute divided by surface area.

If surface area is hard to define, nigh on impossible to measure, and not a particularly good proxy for metabolic rate anyway, why is it still used? The main reason is probably the medical profession's inertia or, more politely, tradition. Medical practice is driven by what works, rather than the best current scientific knowledge. And in truth there is only a small difference in the answers given by calculations based on body area and those based on body weight. Combined with the human body's ability to stabilize itself, adjust to different feeding and fluid regimes, and process different quantities of drugs, the medical use of body surface area calculations, although arguably irrational, does little or no damage. In the same way, the different doses recommended for adults and children on over-the-counter drug labels are not precise calculations, but the approximation should be harmless. Anyway, the issue is becoming moot: In hospitals, direct measurements of patients' metabolic weight will probably supercede calculations based on both body weight and area—provided doctors can be persuaded to give up their time-honored mathematical spells.

DuBois's doubts were in part the result of a long debate with another opponent of the surface law, the physiologist Francis Benedict. By 1930, Benedict had spent two decades measuring metabolic rate in hundreds of humans and other animals. He learned physiology from an American who had shared a lab with Rubner, and in the early twentieth century he extended Rubner's work by showing that alcohol too could be burned to release energy. The work caused consternation in the temperance movement, especially as it was carried out at Wesleyan, a college in Middletown, Connecticut, run on teetotal Methodist lines. Benedict, himself an abstainer, repaid his debt to sobriety during the Prohibition years, when he turned in the chief bootlegger in his Maine hometown to the authorities. In 1907 he left Wesleyan to run the Carnegie Institution's newly established Nutrition Laboratory in Boston.

Benedict is remembered for his work on human metabolism. The Nutrition Laboratory built up a detailed picture of people's energy requirements and how they changed with weight, height, age, and sex.

Along with Eugene DuBois's investigations, the lab set the standard for measuring metabolism: subjects reported for their trial at 8 a.m., having fasted for 12 hours, and lay down for 30 minutes to calm themselves before measurement. Subjects ranged from newborns only two hours old to 93-year-olds who had lived through the Civil War, and from college athletes to vegetarians—who at the time were so rare that Benedict traveled to the Kellogg sanatorium in Michigan to find 10 of them to measure. In 1919 he and his colleague Arthur Harris used this impressive data set to derive equations to calculate metabolic rate from knowledge of height, mass, sex, and age, now called the Harris-Benedict equations. These equations are pretty accurate, although they tend to overestimate metabolic rate by about 5 percent—for me they predict a resting metabolic rate of 1,815 calories per day. They are still widely used to calculate approximate energy needs.

The Nutrition Laboratory moved on from humans to probe the metabolisms of a menagerie of other animals. Besides domestic species such as sheep and cows, and laboratory favorites such as rats and dogs, Benedict persuaded zoos and circuses to lend him exotic beasts such as the cassowary, a flightless bird from Papua New Guinea; the marmot, an alpine rodent; macaque monkeys; and chimpanzees. Having all these measurements done in the same lab, under the direction of one person, gave them an internal consistency that made them the best metabolic comparisons between species done to that date.

In 1938, a year after he retired, Benedict published a monograph, summing up his life's work, titled *Vital Energetics*. The book included a figure showing the increasingly familiar logarithmic plot of metabolic rate against mass, with, as he wrote, "a most gratifying straight-line relationship between the total heat production and the body weight." The gradient of this straight line was 0.73.

Benedict, however, resisted gratification. He saw the pattern in his results as a mirage and warned others against the intoxicating effects of the search for a higher order:

> It is obvious that this apparent straight-line relationship is of no physiological significance, whatever its mathematical significance may be thought to be. . . . It seems illogical to make use of complicated mathematics in the attempt to unravel the end results of the pooled activities of millions of cells, each acting differently. [A]ll attempts by mathematical means to

secure a uniform expression of the basal metabolism findings on different animals species are utterly futile. . . . [N]o unifying principle in metabolism has been found to exist.

It takes an unusual scientist to collect this much information and *not* see any trends in it—most, using the human gift for spotting patterns, will begin joining the dots as soon as they have two data points. And as Kleiber pointed out, measuring the metabolic rate of one animal is a unification, a summing-up of the pooled activities of millions of cells—does that make it a meaningless quantity? "If this is the way Benedict feels," he retorted, "one cannot help but wonder how he ever became interested in conducting a respiration trial." The dispute between lumpers such as Kleiber and splitters such as Benedict about how we should compare different measurements of metabolic rate and what such comparisons reveal continues to this day. But both Kleiber and Benedict agreed that the surface law could no longer stand.

To misquote Thomas Huxley, the surface law was a beautiful theory destroyed by beautiful facts. But Rubner's law was one of those wrong ideas that, because of all the thought, arguments, and experiments it stimulated, proved a lot more useful and influential than many a correct notion. In Rubner's time, biological knowledge, with the giant exception of Darwin's theory, was a pile of facts. Biologists accumulated information about the living world, but there was little attempt to put it into context, to see if any larger structures emerged from the mass of details. In the years between the two world wars, the debate on whether the surface law held or not was one of the most active in biology. Without Rubner's search for laws of metabolism, Kleiber might never have been stimulated to find general trends in the data. Even today, biological principles that make firm predictions across a wide range of species are extremely rare. Kleiber's rule is one of the few examples and one of the most precise. If Bergmann's rule was like Kleiber's, rather than being a general trend with many exceptions, we could point to a spot on the map and say with reasonable confidence what size animals lived there.

Yet Kleiber's discovery only made things more puzzling. To believe that metabolic rate corresponds to the relationship between body mass and surface area, or mass raised to the power of 2/3, is intuitively satisfying and makes good mathematical sense. There is no reason to expect metabolic rate to correspond to body weight raised to the power of 3/4. It's like trying to understand the ultimate question of life, the universe, and everything: If 3/4 (or 42) is the answer, what's the question?

4   SEARCHING FOR SIMILARITY

U NFORTUNATELY, NOT EVERYONE kept up with developments in the biology of body size.

Elephants are intelligent, long-lived animals with complex social and emotional lives. We often flatter ourselves that their mental processes must be a lot like ours. But Louis Jolyon West and Chester Pierce were baffled by one aspect of elephant behavior: musth. A male elephant in musth becomes violent and temperamental, liable to attack anything, human or elephant, that gets in his way. The animal becomes watery-eyed and produces a foul-smelling secretion, brown and sticky, from a gland just behind his eye. Sexually mature male elephants are like this for several weeks each year. To humans in charge of elephants, whether in zoos, wildlife preserves, or as working animals, musth is a hazardous and unwelcome episode. In the United States, safety concerns have led to several captive animals in musth being killed.

To West and Pierce, two psychiatrists at the University of Oklahoma in Oklahoma City, musth seemed like madness—a pointless state in wild elephants and a dangerous one for captive animals and their handlers. It was the early 1960s, and the hallucinogen LSD, lysergic acid diethylamide, was a recent addition to the psychopharmacologist's tool chest. To investigate musth, the researchers decided to manipulate elephant behavior with LSD, which seemed to send people and other animals mad. The psychiatrists teamed up with Warren Thomas, a veterinarian at Lincoln Park Zoo, to test their ideas on Tusko, the zoo's male Asian elephant. If musth was madness, perhaps a dose of LSD would send Tusko into musth. Perhaps it would point to a "cure" for animals of their affliction. Perhaps Tusko would even produce that brown and smelly gunk.

In humans an oral dose of 0.1 milligrams of LSD induces several hours of delirium. Higher doses, of a milligram or more, have powerful physical effects, including raised blood pressure and body temperature, sweating, and dilated pupils, along with psychosis and the drug's trademark visions. But other animals, such as cats and monkeys, seem to be much less sensitive to LSD. Relative to body weight, the dose needed to produce a similar effect in a macaque monkey or a cat was 10 or more times that needed for a human. The psychiatrists decided to inject Tusko with 0.1 milligrams of LSD for every kilogram of his body weight. Since he weighed a shade under 3 tonnes, this worked out to 297 milligrams. Compared with the doses needed to affect cats or monkeys, this amount seemed conservative. "We considered that we were unlikely to see much reaction with this dosage of LSD," they wrote.

At 8:00 in the morning on August 3, 1962, someone at the zoo shot the 297 milligrams of LSD into Tusko's rump with a dart from an air rifle. West, Pierce, and Thomas described what happened next in what must be one of the oddest papers ever published by the venerable American journal *Science*:

> Tusko began trumpeting and rushing around the pen. . . . His restlessness appeared to increase for 3 minutes after the injection; then he stopped running and showed signs of marked incoordination. His mate (Judy, a 15-year-old female) approached him and appeared to attempt to support him. He began to sway, his hindquarters buckled, and it became increasingly difficult for him to remain upright. Five minutes after the injection he trumpeted, collapsed, fell heavily onto his right side, defecated and went into *status epilepticus*.

Tusko lay on his right side, shuddering, his eyes askew and pupils dilated, his left legs stretched out and his right legs curled up. His breathing was labored and his tongue, which he had bitten, had turned blue. The researchers administered massive doses of anticonvulsant drugs but to no effect. At 10:40 a.m., an hour and 40 minutes after the dart was fired, the elephant died. An autopsy revealed that Tusko's larynx had gone into spasm, blocking his windpipe and strangling him.

Poor Tusko, and poor Judy. Even so, the researchers were unwilling to simply write off Tusko's death as an experiment gone wrong. Perhaps, they speculated, the natural musth-triggering hormone was chemically similar to LSD. And perhaps some good could come from Tusko's demise. "It appears that the elephant is highly sensitive to the effects of LSD—a finding which may prove to be valuable in elephant-control work in Africa," they concluded. No conservation organization has pursued this suggestion. (West, who died in 1999, went on to become one of America's most eminent psychiatrists. An expert on cults and brainwashing, he examined Jack Ruby after Ruby shot Lee Harvey Oswald and also Patty Hearst, the heiress-turned-terrorist.)

Yet an understanding of the effects of body size on biology might have saved the trio their place in scientific infamy. You might need a tenth of a milligram of LSD per kilogram of body weight to give a cat a psychotic episode, but to expand Tusko's consciousness, the researchers were wrong to simply give him a proportionate dose, the biggest single (recorded) hit of LSD ever administered—enough to trip out about 1,500 people. Because larger animals have proportionately much slower metabolisms, their cells degrade drugs more slowly, and so they need proportionately smaller doses.

Using a calculation based on the relative metabolic rates, rather than the body sizes, of elephants and cats, an effective dose of LSD for an elephant would be 80 milligrams. Injected with nearly four times this amount, small wonder Tusko got the bad trip to end them all. If Tusko's dose had been extrapolated from the amount typically given to humans, it would have been less than 10 milligrams, even without adjusting for metabolic rate. The knowledge that human metabolisms burn more slowly than those of the smaller animals used in experiments has now percolated through the scientific community more widely than it had by 1962, and today body size is routinely considered in drug trials and dosage calculations on humans and animals.

Tusko's story is a gaudy and catastrophic footnote in the history of biological scaling. But over a half-century of experiments ranging from the quirky to the grisly, researchers' efforts to work out what controlled an animal's metabolic rate, and where Kleiber's rule came from, were just as inconclusive.

Inside or Out?

Max Rubner originally thought that temperature sensors in an animal's skin controlled its metabolic rate, stoking the fires in cold weather. Each sensor controlled a small fraction of total metabolism, he reasoned, so as an animal's area increased, so would its number of sensors and its metabolic rate. This turned out not to be true, however, and eventually Rubner changed his mind and concluded that an animal's metabolic rate was instead an intrinsic part of its biology.

Much of the earliest thinking about metabolism focused, like Rubner, on the regulation of body temperature. But in the early twentieth century, biologists stopped believing that an animal's metabolic rate was a reflection of how hard it was working to keep warm. After all, if that were so, the rate would fall as the animal got hotter, and this doesn't happen. By the time Kleiber, DuBois, and Benedict were working on the problem, the belief that an animal's basal metabolic rate was something that came from within was widespread. Researchers thought that most of the heat produced by animals was a sort of waste product, a thermal background hum emanating from all the body's chemical reactions and physical activity, like the low buzzing noise a fridge makes. From measuring the volume of this metabolic hum, the next step was to locate its source and work out what controlled its volume.

Biologists were searching for similarity: They wanted to find some feature of animals that varied with body size in the same way as metabolic rate. Rubner had thought that this feature was surface area; as his theory fell from favor, scientists began casting around for something to replace it.

Having been let down by animals' outsides, biologists went looking inside, dismantling their subjects into tissues and organs and measuring each component's energy consumption. This promised to show whether an animal's metabolic rate is a product of its parts or its whole: whether mouse cells are always hungry for energy, even when they are not part of a mouse and whether cow organs are inherently sluggish.

In the 1920s, separate French and German teams measured the rate at which slices of tissue from organs including brain, kidney, liver, and heart used up oxygen in a laboratory dish. The tissues came from species ranging from mice to cattle. In their home bodies these tissues from different animals would have burned energy at speeds varying by a factor of 10. In isolation they ran at more or less the same speed. Score one for metabolic rate as a property of whole animals, not their individual organs. But another camp claimed the opposite, that the metabolic rates of an animal's parts matched its whole. In 1941, Max Kleiber did an experiment which, he claimed, showed that even in a laboratory dish a slice of sheep liver used energy more slowly than a slice of mouse liver. As with surface area measurements, the use of different species and experimental techniques made it difficult to compare the results coming out of different labs. How you treat a slice of tissue—what chemicals you bathe it in, for example—has a big effect on how much oxygen it consumes.

Also like the surface law, the debate puttered on for decades. Probably the best experiments were done in 1950 by Hans Krebs at the University of Sheffield in England. Krebs, who won a Nobel Prize in 1953 for unraveling the chemical reactions that turn food into energy inside cells, tested a range of different tissues from different species under different conditions. He concluded that tissues from larger animals had slower metabolic rates but that the deceleration was too small to account for the pattern seen in whole animals.

A logical, if gruesome, conclusion to this line of inquiry was to look at all the organs of an individual animal at once, to see if the metabolic rate of its parts summed to that of its whole. In 1955, Arthur Martin and Frederick Fuhrman, colleagues of Kleiber's at Davis, did just that. They dissected 17 dogs into 23 different organs and measured the metabolic rate of each organ. They did the same for 15 mice, except they were able to extract only 21 organs (dogs were divided into testes and bladder; in mice they left the urogenital system intact). Martin and Fuhrman came down in favor of Kleiber's earlier experiment, and on the opposite side of Krebs, concluding that a disembodied organ had the same metabolic rate as the body it came from.

Rather than coming to any conclusions, this line of inquiry fizzled out. The experiments were too labor-intensive and difficult to repeat and interpret. Again, some of the methods were questionable. Today, laboratories use animals whose origins and biology are well understood. Often the beasts are inbred for generations to make them genetically uniform. Martin and Fuhrman, on the other hand, got their dogs from the local pound.

Maybe a still more detailed view would solve the problem. Metabolic rate, after all, depends on what happens in cells. And a small animal's cells do have more metabolic equipment. They have more of the energy generators called mitochondria and higher concentrations of the enzymes that turn food into fuel, so they use energy more quickly. The concentrations of both mitochondria and enzymes fall with total body size in proportion with mass raised to the power of $-1/4$, identical to the rate at which relative metabolic rate slows with size. This, you could say, explains why relative metabolic rate slows with size, but it really just shifts the problem to explaining why the amount of energy-burning equipment declines.

And disembodiment does have an effect. Cells and organs are sensitive to their surroundings. Many of the experimenters who studied isolated organs tried to take their measurements as soon as possible after the tissue came out of the animal. The metabolic rates of these organs would have carried a hangover from their previous surroundings. But if you take some cells from an animal and culture them in a Petri dish, they make more mitochondria and enzymes and their metabolic rate rises. It's as if they think they have become part of some smaller organism. Over time, the metabolic rates of cells from different-sized animals growing in a Petri dish converge on the same high level.

A cell's energy consumption depends on how many cells surround it. A natural example occurs every time a baby mammal leaves its mother's womb. In the womb a fetus has the same metabolic rate as its mother—it behaves as if it were one of her organs. Pregnant women are advised to eat 300 extra kilocalories per day, enough to both feed their developing baby and increase their own weight by about 14 kilograms (30 pounds) during pregnancy. But when a human baby is born, its metabolic rate rises rapidly. A baby weighing 3.5 kilograms (7.5 pounds) needs about 400 kilocalories a day on its own. It would be foolish to push the comparison too far. Unlike cells in dishes, babies have sophisticated sensory equipment, brains, and a battery of nerves and hormones with which to fine-tune their metabolic responses. But it does suggest that an animal's metabolic rate is more than the sum of its parts.

Around the same time that Kleiber, Krebs, Martin, and Fuhrman were cutting animals up, other biologists were arguing that small animals' bodies weren't made of different stuff than those of large ones, but from different amounts of stuff. As bodies got bigger, their composition should change so that the quantity of what was called "active protoplasm" declined. The most metabolically active tissues, such as brain, heart, and liver, would represent a smaller percentage of the whole, and drowsy fat and bone a greater proportion. As we saw, big animals do have proportionately more bone. And big animals also have relatively smaller brains. For mammals, relative brain mass declines with size: A rat's brain accounts for about 1.5 percent of its body weight, an elephant's only about a tenth of this. But other organs do not change their relative size. The heart accounts for about 0.6 percent of the body mass in all mammals. Nor do small animals power their more rapid metabolisms with relatively larger lungs. Nor do they have proportionately more blood to carry oxygen. Obesity researchers are investigating the relationship between body composition and metabolism. Studies on humans have shown that large people do indeed have less metabolically active tissue, but also that their energy-hungry tissues—basically the major organs—consume less energy, pound for pound, than the same tissues in smaller people. So simple changes in tissue proportions seem not to account for the relative decline in metabolic rate.

Fudge Factors

Another idea, proposed by several biologists, was that Kleiber's rule of body size and metabolism could be a sort of average of all the body's scaling rates, the end result of several different rules acting at once. If

cells in different-sized animals used energy at the same speed, metabolic rate would scale simply with body mass. Tusko would not
have overdosed, but he probably would have starved, as he would have
needed to eat at the same rate as a shrew. On the other hand, cells and
animals must avoid overheating, and they can only do so through their
surface areas, which are proportional to their body mass raised to the
power of two-thirds. Kleiber pointed out that if a steer had the metabolic rate of a mouse, its body surface would be hotter than the boiling
point, and a mouse with the energy consumption of a steer would need
fur 20 centimeters thick to keep warm. Perhaps Kleiber's rule is the
result of a balance between these two forces: 3/4 lies between the
1 predicted by the maximum metabolic rate and cell number and the
2/3 predicted by the surface law.

A similar suggestion was that the 3/4 arose from a balance between
the surface law and the force of gravity. As we saw, gravity places a
greater burden on large animals. Biologists can change gravity: Raising
animals in a centrifuge, a device like a merry-go-round for rodents,
increases the force on them, and firing them into space eliminates it.
The results have been mixed. After two weeks of spinning around and
around, rats consume more oxygen, supporting the idea that gravity
influences metabolic rate. But another study found something like the
opposite: Rats living at G-forces ranging from twice Earth's gravity to
weightlessness on the Space Shuttle seemed to increase their metabolic
rates as gravity's pull weakened. These experiments give no evidence
for gravity as a general explanation for Kleiber's rule—and they were
not designed to, being more concerned with finding out what gravity
does to the body, often with a view toward predicting the medical
hazards of space travel. Another, more damning strike against this
argument is that aquatic animals, buoyed by water, are effectively
weightless, and so their metabolic rates should follow the surface law.
Yet seals and porpoises follow Kleiber's rule just as closely as their land-
living counterparts.

Kleiber thought that some average of scaling effects was the
strongest contender to explain his rule. But this hypothesis never
attracted much support. The reasoning behind the ideas about averag-
ing is hand-waving, working backward from 3/4 by trying out sums

until we get the right number. Ideally, we would like to work forward toward scaling laws, without any preconceived pattern in mind—scientists prefer elegant laws over post hoc fudges. In fact, Kleiber put most of his efforts into arguing for his rule's usefulness and validity, rather than trying to explain it. In his 1961 book on energy in biology, *The Fire of Life*, he spends most of the chapter on the relationship between metabolic rate and body size criticizing the surface law, which by then had nearly expired, rather than presenting any alternative theory. Kleiber was not mathematically inclined and seems to have seen his law as a useful way to work out how much to feed an animal, rather than anything that demanded theoretical explanation.

Biologists tried to pin metabolism down in ever more detail—dissecting bodies, organs, tissues, and cells—hoping an explanation for the trends might pop out of the ever-decreasing circles of precision. Such measurements are doubtless challenging, but there's something unsatisfying about this path. A better description is not an explanation. Even if changing body composition, or an average of different rates, did account for the trend in metabolic rate, it still wouldn't tell us why. Why does metabolic rate seem to be proportional to the 3/4 power of body mass? Where does 3/4 come from? In physics, when an experiment produces something unexpected, the theorists' juices start flowing. In the science of metabolism, strange measurements led mostly to more measurements.

The Metabolic Ark

And my, how biologists measured. As the decades passed, ever more species lined up to have their metabolisms recorded. We have seen that Kleiber's rule applies to both mice and elephants, but this range covers only a small number of species and a narrow spectrum of life. All the investigations considered until now have dealt with mammals—moreover, only the placental mammals like us, accounting for a mere 4,000 of Earth's several million animal species. What metabolic rates did Noah assign to the rest of the animal kingdom?

Marsupials—kangaroos, koalas, possums, and so on—have a lower body temperature than placental mammals: 34°C to 36°C, compared

with 36°C to 38°C. They also, it turns out, have relatively slower metabolisms. A marsupial burns energy at two-thirds the speed of a placental mammal of the same size. But for species ranging from the
9-gram marsupial mouse to a kangaroo weighing more than 50 kilograms, the relationship between metabolic rate and size fits nicely on
Kleiber's line. Birds, the other group of animals besides mammals that
use metabolism to maintain a constant body temperature, have higher
body temperatures than us: 39°C to 41°C. They have faster relative
metabolic rates, too, but again, there is good evidence that metabolic
rate changes with body size according to Kleiber's rule.

So Kleiber's rule seems to hold for warm-blooded vertebrates.
Some mammal species, however, lie away from Klieber's line. This
doesn't mean that we should ditch the whole idea. Allometry's power
lies partly in its ability to highlight exceptional cases—the individuals
or species that deviate from a general pattern—and challenge us to
explain them. There are plausible arguments as to why the metabolic
rates of many of the unusual species might buck the trend. Marine
mammals such as seals and whales use a relatively large amount of
energy for their size, perhaps to keep warm in the sea. Arctic species
also have higher metabolic rates than tropical animals of the same size.
Desert animals, in contrast, burn calories relatively slowly, which could
point to an adaptation to the shortage of food in their environment.
The same goes for animals living in deep or cold water and in caves.
Sloths, too, have slow metabolisms, probably because they are so, well,
slothful. Kleiber's is not an iron rule—animals can bend it.

What about reptiles, fish, amphibians, insects, and other invertebrates, whose body temperatures fluctuate with the ambient temperature? The evidence is not as good as for mammals. The resting
metabolic rate of cold-blooded species (technically known as ectotherms, or poikilotherms) is harder to define than it is for birds and
mammals—it changes much more with temperature, for example—
and studies have estimated wildly different equations relating mass to
metabolic rate. There have also been fewer studies, as these animals are
not raised on farms, so there is less economic incentive to study them,
and they are not such good models for human biology, so there is less

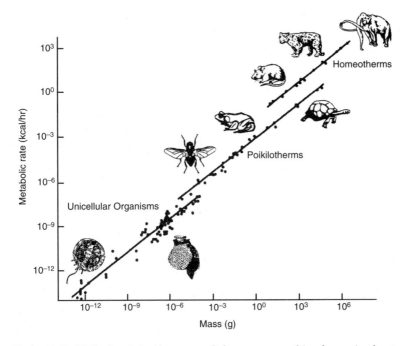

By the 1960s, Kleiber's rule had been extended to cover everything from microbes to monsters.
Credit: Novo Nordisk, Inc.

medical incentive. But one conclusive finding is that a reptile or a fish has a much slower metabolism than a bird or a mammal of the same size, and so needs less food. This is hardly surprising—the ectotherm doesn't need fuel to maintain its body temperature. Big ectotherms also have slower relative metabolic rates than small ones, and over a large size range and a large number of measurements the metabolic rates again converge on the line of body mass raised to the power of 3/4. Still other studies have showed that Kleiber's rule even applied to single-celled microscopic organisms such as amoebas.

This body of evidence contains a lot of variation. Energy has been found to scale with powers of body weight varying from about 1/2 to about 1 in groups from mammals to microbes. Some biologists believe that no unifying trend in metabolic rate exists and that each species' energy needs are the result of its particular physical proportions, chemical makeup, evolutionary history, and environment. Nevertheless, most measurements are close to Klieber's 3/4, particularly if you look at a large number of species across a large range of body sizes— allowing the big picture to emerge from the fog. By 1960, species ranging in size from elephants weighing several thousand kilograms down to microbes weighing less than a millionth of a millionth of a gram, or about 20 orders of magnitude, had all found a place on Kleiber's line.

Quarters, Quarters, Everywhere

It wasn't just metabolic rate that caught the attention of allometry seekers. Biologists got to measuring and comparing just about everything they could. As the power law equations piled up, a pattern began to emerge. Other biological processes, such as mammals' dietary requirements for nitrogen and vitamins, and the rate at which they produce urine, also scale to body mass to the power of 3/4. An animal's heart rate is proportional to its body mass to the power of −1/4, the same as the amount of fuel burned per cell. For every 1,000-fold increase in body mass there is a nearly sixfold decrease in heart rate. Many biological times, such as life span, time spent in the womb, and time between birth and maturity were found to be proportional to body mass raised to the power of 1/4, so a 1,000-fold increase in body mass leads to a sixfold increase in life span. Walking speed also increases with the 1/4 power of body mass. Similar rules apply to biological structures. The cross-sectional area of the aorta, the largest blood vessel, scales as the 3/4 power of body mass. Plants joined in the fun: The area of tree trunks scales in the same way as that of aortas. Other features abandon 1/4 but stick with multiples of it; for example, hemoglobin's ability to seize hold of oxygen declines slowly as animals get larger, with an exponent of −1/12.

The number 4—in 3/4, 1/4, −1/4—or multiples of 4, such as in the case of hemoglobin, came up again and again. It is not surprising that scaling laws should be related, because the different bits of an animal all interact with one another. The demand for nutrients, and the rate at which the body processes and gets rid of them, depends on metabolic activity. The heart rate needs to match the rate at which the cells burn oxygen. In combination the scaling laws display a kind of unity. For example, because metabolic rate for each cell decreases as the $-1/4$ power of body size, but life span increases as the $1/4$ power, each cell burns approximately the same amount of energy in its life, regardless of the animal it lives in. What was harder to explain was why scaling laws should all revolve around the number 4.

By the mid-1960s, most biologists believed that Kleiber's rule was an accurate reflection of the relationship between mass and metabolism. We know this because they voted on the question. At an international symposium on energy metabolism held in the Scottish town of Troon in May 1964, Kleiber argued that 0.75 should be adopted as the standard for calculating metabolic rate from body weight. Despite the objections of those who pointed out that metabolic rate varied greatly within species, between species, and at different times in an animal's life, the conference adopted Kleiber's motion by 29 votes to nil. Kleiber's victory reversed a defeat 30 years earlier, when the same conference had voted in favor of Brody's suggested 0.73. This seems like an odd way for scientists to settle their issues—you would hope that evidence, rather than majority opinion, would be the arbiter. Some of the arguments about Kleiber's rule have a strong whiff of numerology—Kleiber and Brody had fierce arguments about what the third decimal place of the scaling factor should be, even though there was no way of measuring this number precisely.

Agricultural researchers such as Kleiber and Brody needed to know how metabolic rate changed with size, but they weren't much interested in explaining why. Kleiber's rule, and other allometries, gave good predictions, and highlighted exceptions, without any theoretical underpinning, so most biologists probably saw little use for one. A serious mathematical attack on the problem would require someone with a different background, and a fresh angle, to the nutritionists and physiologists who until then had dominated the study of metabolism. This person arrived in the early 1970s.

Intellect and Muscle

Scaling is a large part of biomechanics, the discipline that uses physics to explain how animals move and support themselves. Just as the structure of animals needs to change with size, with bigger animals and plants needing more support, the way they move changes too. Small birds hop along the ground; for any bird larger than a robin, it is more efficient to walk. The largest birds are flightless, because the power of muscles increases more slowly than their weight, so a big bird can never have enough flight muscle to lift itself. A recent study argued that *Tyrannosaurus rex* was too big to run, because it would have needed unfeasibly large muscles to power its massive legs. This might be bad news for the makers of *Jurassic Park*, but it wasn't necessarily good news for the herbivorous dinosaurs of the Cretaceous, as a brisk walk would have carried *T. rex* along at more than 20 kilometers per hour, the speed of a charging rhino. For very small animals, gravity becomes irrelevant, and friction takes over as the dominant physical force in their lives. This is why insects and spiders can run up walls and across ceilings and why the smallest insects have wings like brushes rather than paddles: For them the air is like treacle.

Thomas McMahon trained as a physicist. His doctorate, taken at the Massachusetts Institute of Technology, was in fluid mechanics. His move into biology came via rowing. McMahon, a keen rower, came up with a physical model to explain why rowing shells powered by eight oarsmen or -women go faster than those carrying a single sculler and how much faster they go. This sparked a general interest in how animals moved, and McMahon ended up holding chairs in both biology and applied mechanics at Harvard. There he made great strides in working out the physics of walking, showing how the legs act like two pendulums—something D'Arcy Thompson also believed. The standing leg pivots around the ground, and the swinging one pivots around the hip. He put his discoveries to practical use, designing a running track for Harvard, which, by matching the stiffness of the track to the stiffness of the human leg, reduced both times and injuries. McMahon also had a broad playful streak. He went everywhere, even to official Harvard functions, with his golden retriever. And he published four novels, including one that invented an alternative life for Gordon McKay, the nineteenth-century engineer who endowed his Harvard chair in mechanics. The real-life McKay invented shoe-making machinery; McMahon sent his fictional McKay to Kansas to found a city powered by bees.

McMahon explained Kleiber's rule using a combination of mechanical and metabolic reasoning. Like Karl Meeh, he considered animals as an assortment of interconnected cylinders—limbs, torso, and neck. But whereas Meeh was interested in these cylinders' surface area, McMahon was interested in their mass and strength. His theory revolved around the need for animal bodies to guard against the threat of buckling. Slender limbs make sense: They use less material and are lighter and more mobile. But although a column, such as a leg or torso, might be able to stand firm if it remains vertical, a small bend can have a catastrophic effect, causing the column to fracture and collapse. The elasticity of bone and muscle can guard against this danger to an extent, making the column spring back to the vertical if it bends, but the risk of buckling still prevents bodies from being too skinny. It's obvious that a long thick column is less likely to collapse than a long thin one.

McMahon calculated the rate at which a limb would need to get wider, to protect it against buckling as it got longer. He found that, to avoid such buckling, the width of a limb or torso should be proportional to body weight raised to the power of 3/8. This looks promising—it's an allometry equation and, even better, it's a 1/4 power law. To test his idea, McMahon reanalyzed the data on the dimensions of cattle collected decades earlier by Samuel Brody's team. The girth of a cow's chest—treating the torso as the body's central pillar—was indeed very close to the body mass raised to the power of 3/8. To get from an animal's geometry to its metabolism, McMahon calculated the power that the muscles around such a column could generate. A muscle's power is proportional to its cross-sectional area, which is proportional to the square of its diameter. So to avoid buckling, a body would need a quantity of muscle proportional to its mass, raised to the power of 3/8, and squared. This equals mass to the power of 3/4.

Bingo! Where Rubner thought that animals were similar in the amount of heat their surfaces produced, McMahon thought each animal was similar in the way its limbs and muscles resisted buckling. He called his theory elastic similarity.

McMahon's theory renewed interest in Kleiber's law. But although his ideas worked well for four-legged land dwellers, it's harder to see why the same rules regarding buckling bodies and bending limbs would apply to whales, or birds, or amoebas. None of these animals face the same mechanical problems, but all show 3/4 power scaling. Other researchers argued that the relationship between an animal's size and its shape did not meet the predictions of McMahon's theory and that big animals could get much longer and still not collapse under their own weight than his model allowed for.

Dissenting Voices

Meanwhile, of course, the 1964 vote in Troon hadn't really changed anyone's opinion or settled the question of whether Kleiber's rule was valid. A leading skeptic was Alfred Heusner, who, while working toward his Ph.D. at the University of Strasbourg in France, had discovered that rats show a daily rhythm in metabolic rate. A rat's metabolism runs fastest around the times when it is normally most active, which is at night. Many other species show a similar match between activity cycles and metabolic rate. In 1967, Heusner became a colleague of Kleiber's at what was by then the University of California, Davis. The two were already close friends, and their families holidayed together.

To recap the allometry equation: $y = ax^b$. So far we have fretted about the term describing the power in the metabolic power law, b (also called the exponent), which for mass and metabolism seems to be 3/4 and describes the gradient of the line. Heusner turned his attention to the constant at the front, a, where the line crosses the vertical y axis. He looked at studies of mice, rats, cats, dogs, sheep, and cattle. But instead of lumping all the animals in one species into one group, plotting an average value for each species, and drawing a line through the dots, he considered the species separately. Within a species, he argued, metabolic rate was best described by the old value of mass to the power of 2/3. It was only between species that 3/4 emerged, but Heusner argued that this was biologically meaningless. You weren't comparing like with like. Heusner saw the regularity of the mouse-to-elephant line as a mirage. The one line, with a gradient of 3/4, he said, is really a host of parallel lines, each with the gradient 2/3. It was the differences between these lines, and species, represented by the differing values of the constant a in the allometry equation, that truly demanded an explanation.

Heusner's criticisms were themselves soon under fire, as biologists rode to defend Kleiber's rule. Some said Heusner's mathematical reasoning didn't hold up. McMahon entered the fray, along with his Harvard colleague Henry Feldman. They recrunched the same numbers Heusner had used and concluded that he was wrong to dismiss 3/4 as just an artifact of the data analysis. Two-thirds, they said, might work within a species, where individuals maintained the same proportions as they grew, but the 3/4 seen between species was still meaningful. It showed that the bodies' design did need to change over large size ranges. The trend between species still needed explaining.

The argument between Heusner and Feldman and McMahon set the tone for a debate around Kleiber's rule that continues to this day. Every few months someone unveils a new data set or analysis that, they say, settles the matter in favor of 3/4, 2/3, or neither. Or they claim to be able to demonstrate once and for all that the other guy is talking through his hat. To an outsider everyone seems convincing, or at least convinced. The debate rumbles on, however, and scientists' ability to draw opposite conclusions from the same data seems baffling. In the three-quarters of a century since Kleiber, the literature on this topic has become rather like the Bible: You can find something to support every possible stance. There are wearisome echoes of the arguments we encountered about the surface law and organ metabolism. But, as we've seen, those earlier debates about the surface law and organ metabolism did eventually result in consensus—even if it was a battle of attrition, rather than a coup d'état—and we might expect researchers to eventually come to a broad agreement.

So what will history decide? Employing the Troon principle of majority rule, most biologists believe that Kleiber's rule is a good description of nature. But is it just a handy rule of thumb for farmers and physicians, or is it pointing to an underlying unity of life? Well, for a start, living organisms *do* have an underlying unity. Whales are 10^{21} times bigger than bacteria, but they are made from much the same stuff and sustain themselves in much the same ways, so we might expect to find principles that apply to both creatures. And because the diversity of organisms' sizes is one of the most striking aspects of the diversity of life, we also shouldn't be surprised to find that the patterns in biology are based around changes in size.

On the other hand, organisms' metabolisms depend on a lot of things besides their size. Two animals could be the same size, but their differences in diet, ways of moving about, or breeding behavior, to name a few, could result in very different energy requirements. The high variability between organisms, and each species' idiosyncratic ecological circumstances and evolutionary history, make it hard to spot regularity. Indeed, it means that looked at over small size ranges we should not expect regularity. But if underlying trends are present, we would expect them to emerge over a wide size range, which gives large-scale order a chance to drown out small-scale noise—and this is indeed what we see. This sounds like special pleading, but the same is true in many physical systems. Some quantum effects make themselves apparent only at unimaginably high energies, and some relativistic effects become apparent only over vast distances. Other physical principles are statistical phenomena, applicable to large groups but meaningless when applied to small groups or individuals.

In warm-blooded vertebrates, the organisms whose metabolism is both best studied and easiest to get a handle on, a large body of evidence points to metabolic rate being proportional to mass raised to the power of 3/4. In the 1930s three rival researchers arrived at this conclusion independently. Cold-blooded species are less well studied and their metabolisms are different, but the evidence again points to 3/4 power scaling. Historically, measuring metabolic rate has been difficult, and it is always hard to know how much faith to put into any one data set. But, perhaps most tellingly, even if you ignore metabolic rate and Kleiber's rule, a wide range of quarter power scaling laws are found throughout biological systems. These scaling laws have been found in many situations that can be measured more easily than metabolic rate, such as the width of a blood vessel or the rate of a heartbeat. It is unlikely that everyone who uncovered one of about 200 quarter-power scalings set out to prop up Kleiber's rule.

Animal bodies exist in three dimensions—this is the logic of the surface law. So where does nature's fixation with the number 4 come from? There's one simple way to produce scaling laws that revolve around the number 4. Although we live in a world with three spatial dimensions, it's perfectly acceptable to do geometry on a four-dimensional shape. The proportions of such a shape scale in the same way as those of a three-dimensional one. Just as in a three-dimensional solid, dimension two (surface area) is proportional to dimension three (volume or mass) raised to the power of 2/3, in a four-dimensional shape dimension three is proportional to dimension four raised to the power of 3/4—the very number we are looking for. In 1977, Jacob Blum pointed out that, if you considered living things as having four dimensions, a four-dimensional version of the surface law led simply to a scaling exponent of 3/4. Find an extra dimension from somewhere and all your problems are solved. Blum didn't have any definite ideas about what this fourth dimension might be. Twenty years later, another team of researchers did.

5   NETWORKING

THERE WAS A SAYING going around the Los Alamos National Laboratory in the early 1990s: Biology would be to science in the twenty-first century what physics had been in the twentieth. The researchers at the lab—a physics powerhouse, where the atomic bomb was invented—realized the gathering momentum of fields such as genetic engineering and neuroscience and wondered whether it was biology's turn to change the world the way quantum theory and relativity had.

If biology was truly about to supersede physics as the cradle of scientific revolutions, Geoffrey West had reason to be glum. West had been working at Los Alamos since 1976 and had become head of the lab's particle physics group. But there were signs that the tide was turning against his discipline. At the end of 1993, the U.S. Congress pulled the financial plug on the Superconducting Supercollider, a giant particle accelerator intended to be built near Dallas, Texas. The public, it seemed, was unwilling to spend $11 billion to probe the building blocks of matter. There would be no new accelerators—and so probably no major developments in particle physics—for more than a decade.

The talk of biology's coming preeminence was probably right, thought West. But the idea that this meant physics would go out of fashion struck him as ridiculous. Like D'Arcy Thompson, he saw mathematical descriptions as the root of any true science, and most areas of biology were still lacking in such theories. To West's mind this gave physicists an entry into biology. He began thinking about what sort of biology problems he might tackle.

West had recently turned 50, and his mind had turned to thoughts of mortality. He decided to look into what determined life span. On the one hand, longevity is predictable. Insurance firms and actuaries can calculate life expectancy based on diet, occupation, income, and so on. Half-an-hour's research would have given him a good idea of how much longer he had left. On the other hand, although biologists have some ideas about the mechanisms of aging, no one could explain why a mouse should live for a year or two and a man, built from much the same molecules, with much the same genes and biology, should live for a century. Explanations based in genetics or physiology struck him as superficial. "If biology is to be a real science, you ought to have a theory that can predict why we live 100 years," says West. "That's real science, not some qualitative nonsense about gene expression. That's not an explanation of anything."

He began to teach himself biology in the evenings and on weekends, picking up bits and pieces of knowledge wherever he could. It was, he says, "like learning about sex on the street." The library at Los Alamos is devoted almost entirely to physics, so he resorted to reading his children's high school biology textbooks. He discovered that larger species live longer and that across species life span increases proportional to the 1/4 power of body mass. And he discovered that many researchers had linked life span to metabolic rate, believing that animals that burn energy relatively quickly live shorter lives than the slow burners. But he also discovered that no one could explain why a species' average life span was as long as it was.

West, too, has a background in scaling. Like living things, the behavior of physical systems depends on their size, and the laws of physics are a matter of scale. Quantum theory is most useful for describing what happens inside the atom. Relativity comes into its own when considering interstellar distances and velocities close to light speed. Classical Newtonian physics is good for everything in between, from colliding billiard balls to satellite trajectories. And since the Big Bang, the Universe has been growing. "I viewed the unification of forces, and the origin of the Universe, as a scaling problem," West says. "Since the Universe defines space and time, and it is continually changing its size, what space and time are becomes a scaling problem."

His fumble with the biology literature convinced West that, to understand life span, he needed to understand metabolic rate. To understand metabolic rate, he needed to understand biological scaling. And to understand scaling, he needed to understand the ubiquitous power laws. There was, he realized, a branch of mathematics devoted to just this task—fractal geometry.

Branching into Biology

Crudely put, a fractal is a branching shape that divides and divides, becoming more intricate by producing an ever-larger number of ever-smaller branches akin to the first branch. Since their discovery, such shapes have been associated with natural forms, and many plant and animal structures, such as fern leaves, tree branches, and the air passages of the lungs, look a lot like the fractals drawn by computers, and vice versa. The power of fractals—and a property that suggested nature would use them—is that they use a small amount of information to generate a large amount of complexity. All you need do is repeat a few simple rules: Grow, branch, and shrink; grow, branch, and shrink; grow, branch, and shrink. Instead of encoding an entire shape in their genes—every leaf, every branch and tube—organisms would need just the rules for generating the fractal, in the same way that the equation for a spiral describes the shape of shells and horns.

This repetitive rule means that every bit of a fractal looks the same, regardless of how far back you stand from it or how closely you zoom in on it. And if you chop it into bits, like the sorcerer's apprentice, you get new fractals that look like copies of the original. Identical branches and trees appear ad infinitum. This property is called self-similarity, and it is the link between fractals and power laws. Physicists realized that, when a system can be described using power laws, it is a clue that it is made up of many self-similar parts interacting with one another. Like a power law, a fractal is a rule that generates a hierarchical structure and describes how each level in that structure relates to those above and below it.

The mathematics of fractals opened up new ways of thinking about biological scaling, and of connecting what happens at small scales with what happens at large ones. Before the arrival of fractals, structures such as lungs had been described using vague metaphors such as tree- or cloudlike. But fractal geometry provided a way to replace verbal with mathematical descriptions, just as allometry did for simpler biological forms half a century earlier.

West's chain of reasoning went something like this. Scaling laws are power laws. Fractals describe power laws. So, he asked himself, what fractal might lie behind the scaling of metabolic rate? The answer, he decided, was the system of blood vessels, the network of pipes that carry food and oxygen to the cells, and take carbon dioxide and other waste away from them. This system is obviously like a fractal. It branches from one large tube, the aorta that leaves the heart, and becomes a series of smaller blood vessels. Arteries, such as the carotid, which takes blood to the head, or the femoral, which runs down the leg, lead to narrower arterioles, which lead eventually to capillaries, which reach between the cells and deliver their life-giving cargo to the tissues. West defined metabolic rate as the rate at which resources were supplied to the cells, via the blood system, and he reasoned that the scaling law behind metabolic rate—Kleiber's rule—was a consequence of how the geometry of this supply network changed as animal size changed.

Blood vessels start with one wide tube and branch, like a fractal, to create many small vessels.
Credit: Reprinted with permission from *Science*, vol. 272, p. 122. Copyright 1997 by the American Association for the Advancement of Science.

When he started looking at the blood system, West had to leave Euclidean geometric similarity behind. To the trained eye, a cat's femur and a cow's femur are recognizably the same bone. The big animal needs proportionately thicker bones, to resist the stronger pull of gravity, but the basic structure of its skeleton does not change. But unlike bones, you cannot look at a blood vessel and tell what size of animal it came from. A wide tube and a narrow one might be the same artery from different-sized beasts or tubes from different bits of the same animal. Such big and small blood vessels are, however, geometrically similar to one another. Euclidean similarity is lost; self-similarity is gained.

So West tried to build a fractal that described blood vessels and relate that to metabolic rate. This would, he realized, be an abstraction, a sort of cartoon of how animals worked. It would not account for all the complexity and variability seen in real vascular systems, but West thought that, by ignoring the details of particular systems, he could get to deeper principles underlying them. Physics has often worked like this, by discarding much of the detail in the systems it seeks to study, such as friction in mechanics. For example, when Galileo dropped objects off the Leaning Tower of Pisa, he ignored their differences in air resistance and concluded that they all fell at the same speed, thus paving the way for a theory of gravity. West is fond of saying that, had Galileo been a biologist, paying more attention to the details, he would have ended up writing tomes on how every object falls at its own unique speed. He (West, that is, not Galileo) can talk like this because, as a physicist, he is not too concerned about annoying biologists.

So West began building a fractal that described blood vessels, to work out how fluids would flow through such a network and to relate that to metabolic rate. He eventually designed such a network, but it was a sorry creature, even as a cartoon, bearing little resemblance to any real blood vessels. West realized that he didn't understand the biology of blood vessels very well and furthermore that some of his mathematics and physics were mistaken.

As a scientist looking to jump fields, however, he had one big advantage: the Santa Fe Institute. The institute was set up in 1984 by a group of senior researchers at Los Alamos, along with other eminent physicists from across the United States, to address the problems in chaos theory, complex systems, and emergence that were beginning to make an impact across the physical, biological, and social sciences—problems that seemed somehow to combine hideous complexity with tantalizing flashes of order. Such complex dynamics and emergent order were a common feature of any system consisting of many inter-acting parts, be it stock markets, cells, ecosystems, or societies. The institute was intended to be truly multidisciplinary—it has no depart-ments, only researchers. Since then, Santa Fe and complexity theory have become almost synonymous.

Twenty years on, the institute, now situated on a hill on the town's outskirts, must be one of the most fun places to be a scientist. The researchers' offices, and the communal areas they spill into for lunch and impromptu seminars, have picture windows looking out across the mountains and desert. Hiking trails lead out of the car park. In the institute's kitchen, you can eavesdrop on a conversation between a paleontologist, an expert on quantum computing, and a physicist who works on financial markets. A cat and a dog amble down the corridors and in and out of offices. The atmosphere is like a cross between the senior common room of a Cambridge college and one of the West Coast temples of geekdom, such as Google or Pixar.

At the time he began thinking about metabolism, West was a visit-ing fellow at the institute, spending one day a week there. Through Santa Fe, he met two biologists, based 60 miles down I-25 at the Uni-versity of New Mexico in Albuquerque, who had also been thinking about metabolic rate.

Enter the Ecologists

Brian Enquist began his undergraduate education hungry for big unsolved problems that would need big ideas to explain them. The last place he expected to find them was in biology—Darwin seemed to have sewn up the market more than a century before. Enquist thought he might major in philosophy. But then in his initial biology lectures he saw the graph of body mass versus metabolic rate and learned that there were still big patterns in biology that cried out for big ideas to explain them. He got hooked on scaling.

Every biologist has his or her favorite branch of life—birds, bees, moths, water lilies, coral reef fish, or whatever. For Enquist it was trees. Ever since he began climbing them as a child, he had had a thing about them. They were impressive and important. If you stand in a forest, the world seems to be made of trees. He was also impatient, and botany is a good discipline for a naturalist in a hurry, because you don't have to sit in a hide all day waiting for trees to come to you, and it doesn't take a week to get your sample size into double figures. And for a young biologist beginning his career and looking for big questions to answer, botany was fertile ground. The main thrust of the science has been to classify and describe plants, and there is relatively little theory or mathematics in the discipline; new ideas in biology have tended to be developed and explored by researchers working on animals. Enquist decided to investigate scaling in forests, to see how the sizes of individual trees influenced the form of the whole forest.

For graduate school he went to work in Jim Brown's lab at the University of New Mexico. More than a decade before Enquist joined his lab, Brown had decided that energy was the key to understanding biodiversity. He had begun his career in the 1960s as a physiological ecologist, studying the biology of energy in animals, looking at how the challenge of keeping warm affected their energy budgets and how it would affect their behavior and where they could live. One of his early studies was on the metabolic rate of weasels, and the cost of being long and thin, with a relatively large surface area—a weasel burns energy twice as quickly as a round animal of the same weight, he found. Lately, he had come to see energy as an organizing principle for the whole of nature. How organisms got energy and divided it between themselves could, he believed, explain biodiversity: Why different environments contain different numbers of species, why species live where they do, why certain species are found together or not, and why some are common and some are rare.

And the foundation of any investigation into energy and biodiversity, Brown believed, should be body size. Body size controls how much energy plants and animals need and so how much is left over for other individuals and species. Body size is also closely related to virtually everything else that ecologists are interested in, such as how much land animals need, how quickly their populations grow, and how many young they produce.

The ecological importance of body size makes it of practical as well as academic interest. To conserve a species, we need to understand it. As the biologist William Calder wrote: "A conservation biologist trying to prepare a protection plan without data on the species' biology is like an insurance underwriter issuing a policy in ignorance of the applicant's age, family status and medical history." Yet conservationists often work in almost perfect ignorance of what they are trying to save. It is likely that at least 90 percent of species, and possibly a much higher proportion, have not yet been discovered, described, and named. And our knowledge of most of those that have stops with their existence—we have only a pressed leaf or a bug pinned out in a draw. No one knows how widely they are spread, how great their numbers are, what they eat, what parasites assail them, how many young they produce and how often, what foods or habitats they prefer, or how they behave. Of the species we do know something about, the majority are birds, mammals, and flowering plants. Of the most diverse groups, such as insects, fungi, and bacteria, we know practically nothing. Most species live in tropical forests, through which it is difficult to travel and in which it is difficult to spot things. The number of biologists and the resources they have to fund their discovery and description of species are both limited, particularly in tropical countries, which tend to be among the poorer nations. Our destruction of biodiversity outpaces our knowledge of it.

We need rules of thumb that, in the absence of detailed knowledge, can help us predict which species are most vulnerable to threats such as habitat destruction or hunting and so which are most in need of conservation. Rules based on body size have two great advantages. Size is informative, and it is easy to measure. Even if we know nothing else about a species, we almost always know how big it is. It's been said that ecologists report body size in the same way that journalists report their subjects' ages—practically as a reflex. It takes only a moment to weigh an animal; taking such a measurement requires little equipment or expertise, its accuracy can usually be trusted, the animal doesn't have to be alive to be weighed, and if it is, you don't need to interfere with it or detain it for long. Size is the only universal piece of biological data.

One of the most reliable rules of conservation is that it is bad to be big. Across the animal kingdom, big species are more at risk than their smaller brethren. Orangutans and elephants are disappearing more quickly than rats and rabbits. Rhinos are more endangered than antelopes. Large birds, such as albatrosses and eagles, are in more trouble than warblers and finches. Whales are more endangered than porpoises. This rule holds even for reptiles and insects: The planet's largest known earwig, a 3-inch-long giant from the Atlantic island of St. Helena (being an island species is another almost guaranteed recipe for trouble), has not been seen for 20 years.

It's easy to see what makes large species more prone to extinction. They need more food and so more land. They are often found higher up the food chain and so depend on everything below them staying in good shape. All of this means that there are fewer of them to start with. Big animals take longer to reach breeding age, breed less often, and produce smaller numbers of young when they do. The fossil record shows that, throughout history, carnivorous mammals, such as cats and dogs, have experienced a high degree of evolutionary churn. Species come along, dominate for a few million years, and then disappear, at which point a new group comes along to do the same job. There are obvious benefits to a predator in being big and fierce, but this might also paint a species into an evolutionary corner. Populations become smaller and more spread out, and anything that reduces their food supply or splits a population into isolated fragments, as deforestation and urbanization are doing for large carnivores these days, will hit them harder than smaller species with less grandiose diets.

There's no such thing as small-game hunting, and human hunters' lust for large prey has exacerbated big species' vulnerability. Since humans appeared on the scene, a swath of large mammals, such as mammoths, mastodons, and giant ground sloths, have gone extinct. The same goes for the largest birds—the moas of New Zealand, the elephant birds of Madagascar, and perhaps the largest of them all, *Genyornis*, an Australian species that weighed 100 kilograms. When humans arrived in Australia the continent was also home to a lizard,

Megalania, which was 5 meters long and weighed 600 kilograms. This too is no more. The trend continues to this day: Fish stocks of large-bodied, slow-growing species are slower to recover from fishing and so more vulnerable to overfishing. Big charismatic animals are the poster children of conservation. We study them more intensively, bias our conservation efforts in their direction, and value them aesthetically and spiritually, but it hasn't done them much good.

So when we plan where to spend scarce conservation money, working out which species are most likely to be at risk of extinction, there is a good case for biasing our efforts toward large species. And when we find a new species or population, we should be more concerned for its future if it is an ape or a deer than if it is a rodent.

The Insurance Man

When Jim Brown and Brian Enquist began thinking about metabolism, they were more interested in explaining nature than saving it. The search for general principles that apply across the living world, the pair believed, should focus on a combination of energy and allometry. This call to make energy the center of ecology was not unprecedented. Ecology is the study of nature's economy, and energy has long been one of the currencies tracked by ecologists, through food webs, for example, to explain how much life a habitat can support or as a way to understand why animals prefer to eat certain foods. Ecology has also been one of the areas of biology most receptive to ideas from physics. Around the turn of the twentieth century, ecologists seized on concepts then influential in chemistry, such as equilibrium and thermodynamic descriptions of energy flow, and applied them as metaphors for explaining ecological phenomena such as the stability of populations and the coexistence of species.

The person who did the most to get ecologists thinking like physicists was Alfred Lotka, born in Austria in 1880. Lotka did nearly all his scientific work in his spare time. He trained as a physical chemist and emigrated to the United States in 1902. There he worked at the General Chemical Company, in a patent office, at the U.S. Bureau of Standards, and as a science journalist. He ended his working life, following a brief stint in the 1920s as a full-time researcher at Johns Hopkins University in Baltimore, at the Metropolitan Life Insurance Company, where he made seminal contributions in applying mathematics to the study of human populations, such as calculating how mortality changes with age.

Beginning in the early 1900s, Lotka began thinking about how physics might be applied to biology. Like D'Arcy Thompson, he made no distinction between biological and physical systems. But Lotka saw life in terms of the exchange of energy, rather than being governed by physical forces and geometry.

Lotka was not the first person to have this idea, but he pursued it harder and farther, and with greater mathematical rigor, than anyone else. Also like Thompson, he worked outside the academic mainstream and had little contact with other scientists. A quarter of a century of such work culminated in his 1925 book, *Elements of Physical Biology*. In the book Lotka imagines physical biology, which he defines as "the application of physical principles in the study of life-bearing systems," as analogous to the then-voguish physical chemistry. Scarcely any life-bearing system escaped his gaze. He tackled growth and population dynamics, the cycling of chemical elements from the environment into life and back again, behavior, the senses, communication, travel, and consciousness. Often he took examples from economics and sociology. Lotka showed that the sheep population of the United States matches the sheep consumption of its citizens and argued that predators similarly control the numbers of their wild prey. And he speculated on the immense impact that the newfound ability to take nitrogen from the air and turn it into fertilizer would have on humanity and the planet. Lotka wanted to build an intellectual framework that would unify physics, biology, and the study of human society.

As well as trying to understand biology in terms of physical principles, Lotka sought to solve biological problems using the tools of physics. He drew analogies between evolution and thermodynamics, but he also thought that organisms were far too complicated to be understood simply by the application of thermodynamics. It would be "like attempting to study the habits of an elephant by means of a microscope." But the mathematical techniques of physics could still be used for studying life, by treating plants and animals as if they were particles: "What is needed is an altogether new instrument, one that shall envisage the units of a biological population as the established statistical mechanics envisage molecules, atoms and electrons."

The world, said Lotka, is an engine, and although it is useless—all its work is used internally to feed and repair itself—the world engine's great trick is its ability to improve its workings as it goes along, through evolution. He imagined life as being like a water wheel, turned by the energy flowing over it. Natural selection, he said, would make the wheel bigger and make it turn faster, favoring those organisms that were best at grabbing energy and those that used it most quickly. In consequence, the world engine would speed up. The law of evolution was the "law of maximum energy flux," and natural selection was a law of thermodynamics.

Lotka's thinking had a huge influence on ecologists. The *Elements* was a critical point in ecology's long history of using mathematical models to convert verbal arguments about how nature works—"the birth rate will fall in crowded populations," "predators will reduce the numbers of their prey"—into precise and testable statements, turning words into numbers. Now there is a whole subdiscipline devoted to mathematical models and computer simulations: theoretical ecology. The advent of theory has led to tension between some theoreticians, who see their naturalist colleagues as stamp collectors, and some naturalists, who suspect that the theoreticians, lost in their mathematical fairy world, could not tell the difference between a lion and a dandelion.

Thinking Big

As well as sharing an energy-centric worldview, Brown, like Lotka, thought that revealing the unity of nature required a new instrument. Brown had become frustrated with ecology's emphasis on small spatial scales and short periods. The typical experiment tracked, say, the plants in one field over three years or the molluscs on a single beach over a single summer. A great many of the published studies focused on 1 square meter of land or less. Such small-scale experiments revealed the complexity and unpredictability of nature: The result of, say, removing a predator or adding nutrients to a site varies hugely between places and times. But faced with such variability, this approach offered little hope of saying anything universal about the living world. "There was the idea that to understand systems of many interactive components, such as ecological communities, we should take them apart, figure out what the components are and how they work in isolation, put them together in pairs, and then build up more complicated systems. That doesn't seem to have worked," Brown says. You can create a mathematical model of an ecosystem, but as you add variation and complexity, such as more species or different environments, trying to make the model more lifelike, the number of possible outcomes, such as which species will go extinct, which will spread, and how long all this will take, soon becomes astronomical. Tiny variations in conditions and assumptions lead to radically different outcomes. And there is usually no way to tell which path is most likely in the real world and why.

What was needed, Brown thought, was an effort to find out what united and divided environments and species right across the earth. It would be practically impossible and ethically wrong for a scientist to manipulate nature on this scale, but a program of measuring and observation should reveal the big picture. It would be a search for emergence: Just as physicists know how changing the temperature or pressure of a gas will affect its behavior without knowing what every molecule in that gas is up to, so statistical techniques applied to large groups of organisms—treated, as Lotka urged, as if they were particles—should reveal regularities.

Such large-scale observations had already revealed many large-scale trends in nature. There is a regular relationship between the size of a place and the number of species found there. Most species are rare, with small population sizes, while a few are common. Most species live in only a small area of land; a few roam across continents. Most of the higher groups used to classify plants and animals, such as genus (e.g., *Homo*), family (e.g., primates), and order (e.g., mammals), contain a relatively small number of species; a few, such as beetles, are fantastically diverse. And similar mathematics can be used to describe the ratio of rare to common, or small- to large-ranged species, or diverse to less diverse groups. Something seems to be going on that is general but hard to explain.

What Brown *did* invent, along with his colleague Brian Maurer, was a name for this approach—macroecology. In 1995, Brown published a book of the same name. Not everyone agreed with this view of nature. Like West, Brown was willing to sacrifice the detail to see the generalities. "My knowledge of biology is more extensive than intensive, and I got a reputation as being a big-picture guy," he says. "I have a very broad biological background, and a good feeling for how organisms and systems work, at the sacrifice of detail." It's like squinting at a picture so one can see the general patterns without being distracted by the brushwork. But squinting is a tricky thing—people see different pictures, and some see just a blur. Sometimes the eye of faith takes over, and some ecologists suspect Brown of having a mystical streak. There can also be snootiness about science that is based on observation rather than experiment. The results of experimental manipulations—discovering that changing system X using method Y gives result Z—are often seen as more conclusive, more revealing of underlying mechanisms and true causes, and somehow more scientific than statements about nature based on the passive observation of patterns. Then again, no one thinks astronomers are unscientific because they do not tinker with galaxies or geologists because they do not move tectonic plates themselves.

Into the Woods

In the summer of 1995, Brian Enquist took a copy of Brown's *Macroecology* off on a field trip to the forests of Costa Rica. He returned with a question—did plant metabolism also scale to the 3/4 power of body mass? This is the sort of problem that biology graduate students often end up tackling; it's known as the aardvark approach. ("Nobody's looked at digestive enzymes/the genetics of eye color/parental care in the aardvark," says the supervisor to her new student in need of a research project. "Why not do that?") Brown told Enquist that nobody had looked to see if Kleiber's rule held for plants. Enquist decided he would take a look.

The first task was to work out what a plant's metabolic rate is and how to measure it. Plants don't burn energy to keep themselves warm, and they don't take in food—they make their own by photosynthesis. But they do respire, just as animals do, and take in oxygen and give off carbon dioxide, so an experimenter could put a plastic hood over a plant and measure its gas exchange. But there is a problem with this procedure. For a start, trees are big and difficult to seal inside an airtight bubble. More fundamentally, when they respire, plants give off water vapor, just as we do. But unlike us, they have no lungs to pump this water vapor out. The rate at which they lose water from their leaves depends instead on the amount of water in the air, the difference in humidity between their insides and outsides. Evaporation from the leaves draws water in through the roots, from which it flows through the plant in tubes called xylem. Sticking a plastic bag over a plant makes its external environment more humid, which makes it harder for the plant's leaves to lose water and lowers its metabolic rate. In other words, the experiment changes what it seeks to measure.

Instead, Enquist settled on using the rate at which fluid flowed through a plant's xylem as a proxy for its metabolic rate. The rate of respiration, and the rate at which resources were consumed, he reasoned, ought to be proportional to the rate at which water flowed through the plant, because the former depended on the latter. Faster flow should translate into a faster metabolism. Foresters want to know trees' flow rates for the same reason that farmers want to know animals' metabolic rates, because it helps them work out how much water their crop needs and how much wood it will produce. They had already built up a handy number of measurements for different tree species, which they related to plant size.

So Enquist hit the library. After a few weeks of rummaging, he had found enough measurements relating plant fluid flow to size to allow a statistically solid analysis. Pooled, these measurements showed that fluid transport was proportional to the plant's mass raised to the power of 0.733—as near to 3/4 as makes no difference.

Emboldened by this success, Enquist suggested to Brown that they seek a theoretical explanation for Kleiber's rule. Brown had often wondered in passing why scaling rules tended to group around multiples of one-quarter, but he had never mounted an assault on the problem. He agreed to tackle the project but warned Enquist of the question's knotty history. It might take years of work, and there was no guarantee of ending up with anything to show for it.

The initial investigation into trees, and the way they transport water, set Brown and Enquist thinking along the same lines as Geoff West: networks. Like him, the two ecologists thought the answer to the puzzle of metabolic rate might lie in the geometry of the transport networks that move fluids around organisms. All large animals and plants are riddled with such networks. Vertebrates have blood vessels and plants xylem. Insects' bodies are honeycombed by tubes, called tracheas, that carry air to their tissues. Perhaps the scaling of such networks could explain the scaling of metabolic rate. Unlike West, however, they began to design networks using graph theory, a branch of mathematics that deals with the properties of networks of interconnected points and lines. Unfortunately, all of the networks they produced using this technique delivered resources at a rate proportional to the 2/3 power of their size—they had reinvented the surface rule. Brown, a self-confessed "mathematical cripple," realized they needed some help.

All for the Best

In the autumn of 1995, Brown and Enquist met West, through the Santa Fe Institute and a mutual acquaintance. The three realized they were taking a similar approach to the same problem. They also realized that working together would save a lot of time: West could handle the mathematics and physics; Brown and Enquist could tell him if the models made biological sense. As long as they trusted each other's expertise, and as long as they took the time to explain things to one another, none of them would have to spend years blundering through a field about which he knew next to nothing. They began meeting on Fridays to work on networks and metabolism.

At first these sessions were slow going. West despaired of ever communicating with the two ecologists. "There were many times," he says, "when I left Santa Fe and I thought this is just hopeless—they don't understand anything, and I don't know what the hell they're talking about." Likewise, Brown and Enquist had to take the time to explain to West such concepts as how evolution works and what a species is. But after a couple of months, they felt they understood one another. Most importantly, they realized that, despite their different backgrounds, they shared a similar philosophy—that there is a unity underneath the diversity of life—and that they got along well enough to be able to work together.

Brown and Enquist put plants to one side, and all three began working on the vertebrate blood system. Most studies of metabolism have been done on mammals, so explaining their metabolic rate would be a good place to start. They began designing the blood system as if it were an engineering problem. If they started from scratch and tried to design a delivery system to distribute resources around a body, what would the resulting network look like? The way that metabolic rate changed with size, they believed, would reflect the way that these networks changed. (In basing their theory on fluid networks, they were also resurrecting one of biology's oldest metaphors. In the seventeenth century, thinkers such as Descartes saw life's workings in hydraulic terms, as a series of pipes, like the fountains and plumbing of Versailles.)

It might seem as if redefining metabolic rate as the speed with which tubes deliver resources gets things the wrong way around. Metabolism is a property of cells, not networks, and you might expect the network to accommodate itself to the cells' demands, not the cells to adapt to what their suppliers can deliver. In fact, the two should go hand in hand. There's no point in having blood vessels that can pump resources faster than cells can burn them, or in having cells that demand more energy than transport networks can deliver. Each link in the chain should be equally strong. Evolution, faced with many things to do and a finite amount of time, energy, and resources to do them with, must strike a balance, making life a matter of trade-offs. Reinforcing one part of one's biological armory means neglecting another. Things need to be good enough to do their job, but not so spectacularly well constructed that they become cripplingly expensive to build and maintain. You wouldn't waste money building a garden shed to the same specifications as a nuclear bunker; a jalopy will sit in a traffic jam just as well as a sports car; and only an eccentric, when faced with an uncracked nut, would rush out and buy a sledgehammer.

Experiments support the idea that respiration is integrated and efficient in this way. One study looked at the maximum capacity of each part of the respiratory system—lungs, blood vessels, and mito-chondria—in mammals ranging in size from shrews to cows. Remov-ing each component from those up- and downstream, and supplying it with oxygen directly, showed how well each part could perform in isolation—if, say, the flow of oxygen to the blood did not depend on the lungs or if the supply to the mitochondria did not depend on the blood. It turns out that a blood vessel's maximum capacity for carry-ing oxygen matches mitochondria's maximum capacity to burn it. And the mitochondria of all mammals burn energy at the same rate, so metabolic rate depends on the number of mitochondria per cell, rather than the activity of mitochondria.

But West, Brown, and Enquist went beyond assuming that cells and networks were *equally* efficient. They sought to design a network that was *maximally* efficient, delivering the maximum amount of resources using the minimum amount of time and energy. This tech-nique, called optimality modeling, is well established in biology, but it is also controversial. It works by calculating the best possible solution to a problem and then seeing whether reality conforms with that model. To give two examples: Is it better for a shorebird to eat a small clam that is easily pried open or to invest the time and effort in smash-ing open a big mussel with a thick shell? And should a male dung fly stay with his current mate, to prevent her from mating with other males and throwing away his sperm, or will he father more offspring by going off to look for another female? So, unlike an allometry model, when researchers look at nature and then see what mathematical expression best describes what they see, in optimality modeling the researcher asks him- or herself what the best way to do a particular job is, works out the biological consequences, and then tests the hypothesis on real animals. If your theory correctly predicts biological phenomena, you can be more confident that you have understood the processes under-lying the phenomenon.

Optimality modeling has been contentious because, although organisms are clearly good enough at what they do to survive, there is no reason to assume they are as good as they can possibly be. Some biologists accuse optimality theorists of a blind faith that evolution has honed every trait to its keenest possible edge. Such skeptics, notably Stephen Jay Gould and Richard Lewontin, have accused optimality enthusiasts of pushing a "Panglossian paradigm," after the character in Voltaire's *Candide* who believed that everything must be for the best because this was the best of all possible worlds. The doubters point out many reasons to think that life is not optimal. You can't just consider one feature in isolation. The aforementioned trade-offs mean that improving one thing, such as lungs, means taking from another, such as legs. Some biological necessities might contradict one another— there's no point in a mouse going after some juicy berries if there's a snake sitting on the same branch (although the berries will appear juicier, and the snake less scary, as the mouse gets hungrier). All organisms are in thrall to their evolutionary past. They cannot rip themselves up and start again, but must deal with current circumstances using what they've got. For example, many aquatic insects carry a bubble of air with them and must return to the surface for refills. Gills might be better, but these insects' ancestors went too far down the air-breathing road to turn back. Evolution might not have caught up with environmental change, forcing organisms to play catch-up. For example, the migratory and breeding patterns of some European birds are now out of synch with the seasons, thanks to climate change, and only a handful of humans have immune systems capable of shaking off the AIDS virus. Finally, natural selection is not necessarily an optimizing process. It has no goal in mind, no notion of what is best.

But while it's unreasonable to assume that plants, animals, or microbes are perfect, it's reasonable to assume that evolution has made them good at what they do. Optimality theory is one way to work out what this means and ask questions about adaptation in precise and testable terms—it is a hypothesis, not an assumption. Used this way, optimality theory has been a useful tool for biologists, and living forms and animal behaviors often turn out to be close to the engineered solution to a problem. For example, a moose's menu consists of land plants, which are high in calories but low in essential sodium, and water plants, which are rich in sodium but bulky and poor in other nutrients. A moose chooses a diet that maximizes its energy intake while eating just enough pondweed to meet its sodium needs. The more important the task or structure, the greater the force that natural selection will have brought to bear on it. An organism's distribution networks are just as crucial as its bones, so while we might not expect them to be as efficient as theoretically possible, optimality is a reasonable thing to look for in a model network. Or as D'Arcy Thompson wrote in *On Growth and Form*:

> That this mechanism is the best possible under all the circumstances of the case, that its work is done with a maximum of efficiency and at a minimum of cost, may not always lie within our range of quantitative demonstration, but to believe it to be so is part of our common faith in the perfection of Nature's handiwork. . . . To prove that it is the very best of all possible modes of transport may be beyond our powers and beyond our needs; but to assume that it is *perfectly economical* is a sound working hypothesis. And by this working hypothesis we seek to understand the form and dimensions of this structure or that, in terms of the work which it has to do.

Plumbing

Besides delivering resources quickly and economically, a well-designed network must also be comprehensive. No part of the organism should be more than about a millimeter—the distance over which diffusion can take over from pumping—from a branch of the network. So West, Brown, and Enquist decided that their ideal network must deliver to every nook and cranny of a three-dimensional space, using as little time and energy in the process as possible. A fractal is a design for transforming a tubular blood vessel into a space-filling solid. After a few rounds of branching and shrinking, the large original tubes have divided into a feathery mass that infiltrates every part of the volume it occupies. Instead of asking how metabolic rate changes with an animal's mass, then, West and company were asking how energy requirements relate to a body's volume, like their predecessors who had studied the surface law.

But natural structures aren't true fractals. If you look at the fronds of a fern under an electron microscope, you don't see atom-sized fern leaves, bearing still more fronds. At some point the branching stops. In the blood system this network terminus is the capillary. From a capillary's viewpoint, animals all look the same. Capillary size varies little between animals; an elephant's capillaries are about the same size as a mouse's. The same goes for cell size—big animals don't have bigger cells, just more of them. So although animals vary hugely in size, the ends of their distribution networks—the bits that plug into the cells— do not. The same goes for plants. The leaves on an oak tree are about the same size as those on a shrub, and the leaf stalks that deliver water to them are the same width. This is squinting at nature. Of course, cell and leaf size do vary—a rhubarb plant's leaves are much larger than a giant sequoia's needles. But the important thing is that leaf size and capillary size do not vary systematically with body size. Large leaves are as likely to be found on small shrubs as they are on giant trees. So when considering the effect of body size, this is one detail that can safely be discarded.

The effect of body size on transportation networks only starts to appear as you begin to travel up the system from the cells. On the reverse journey from the capillaries to the heart, the tubes become wider. It's like journeying up your bathroom faucet, through the pipes of your house's plumbing, and out into a water main. In blood vessels, you eventually reach the aorta leading from the heart. Here, differences in size do matter. Big animals need bigger hearts, as they must pump a larger volume of blood a greater distance, and they need wide tubes, to carry all that pressurized blood with little resistance. A human aorta, at about 2 centimeters wide, is less than a tenth as wide as a sperm whale's—a whale's heart would explode if it tried to pump into a human-sized aorta. So because both whale and human networks end in capillaries of the same width, networks must get to the same end point from different starting points. A bigger animal's blood vessels have got a lot more narrowing to do.

I hope this is beginning to show how body size can affect network structure and so the delivery of resources. West, Brown, and Enquist reasoned that the changing journey from aorta to capillary would determine how metabolic rate changed with size. These changing delivery routes also show how the scaling of networks differs from the scaling of body parts such as bones. Because the networks of different-sized animals have different journeys, their blood vessels cannot be geometrically similar, and the scaling rules relating surface area to volume are no longer useful. In this sense a big animal is not just an enlarged version of a small one. If an elephant's capillaries were thousands of times bigger than a mouse's, they would be useless at delivering resources because their surface area would be tiny compared with their volume. The larger animal would be unable to extract oxygen and nutrients from its blood quickly enough to sustain itself. What an elephant needs, compared with a mouse, is more of the same—more capillaries. But more capillaries means more branches in the network, which means a longer journey from heart to cell, which means a slower supply of resources to the cells, which means a slower metabolic rate. The challenge is to put this in mathematical terms and work out where 3/4 comes from.

So the rule of network design is to start with one big tube and turn it into a three-dimensional fractal, until you have made enough capillary-sized tubes to service your entire body, and then stop. To do this, the tube must branch. The total number of capillaries is the number of branches produced at each junction, multiplied by itself once for every junction traveled through. So an animal with three junctions, whose tubes split into two at each junction would have $2 \times 2 \times 2 = 8$ capillaries. As junctions are added to the network, the number of capillaries rises exponentially, so filling a big animal with capillaries needs only a few more junctions than filling a small one. Again, fractals are the economical solution.

At each level of the network the tubes downstream are narrower and shorter than those that lead into them. To go straight from an aorta-sized tube to a mass of capillary-sized ones would be as disastrous as trying to drain a sperm whale's heart into a human aorta. The turbulence, and resistance, in the tube would shoot up and so would the energy needed to force the fluid down the pipe. What is needed is a series of junctions easing gradually from big tubes into small. West, Brown, and Enquist were like plumbers trying to bolt a bunch of smaller pipes onto one big pipe. They had to choose how much narrower each pipe should be than its predecessor to most efficiently carry the fluid that was gushing down the big pipe. They also had to bear in mind their destination, the mass of capillaries spread out through the body that need to be hooked up to the network. This presented them with another choice: how much space to leave between junctions to keep the total journey as short as possible. In the final network, ease of flow depends on tube width, and journey time depends on tube length.

To work out how much narrower and shorter than its predecessor each successive section of tubing in a perfect network should be, the trio turned to the laws of fluid dynamics. Using fluid dynamic principles, they were able to show that, to minimize congestion in the system, the total number of downstream tubes should have a combined cross-sectional area equal to those upstream of the junction, a property called area-preserving branching. So if the aorta split in two, for example, each offshoot would have half of the aorta's area. The belief that this is how real networks behave goes back centuries: Leonardo da Vinci noticed that the total area of a tree's branches is approximately the same at each level, from trunk, to branches, to twigs. To achieve area-preserving branching throughout the network, the tube's width needs to decrease by a constant proportion at every split. The second requirement, to fill the volume with a minimum length of tubing, is achieved by maintaining a constant ratio of lengths between each junction— each pipe might be half or a third as long as its predecessor, for example.

Let us return to the whole network and sum up its assembly instructions. Start with a single fat tube of a certain length and split it into a number of smaller tubes that are scale models of the original. Split the daughter tubes again, into another generation of tubes that are scale models of their parents, using the same proportions as the first split. (The exact proportion by which each tube should get shorter and narrower turns out to depend on the number of branches added at each junction.) Repeat until you reach the capillaries.

Here we run up against another trade-off. More capillaries ought to be better. More capillaries means more resources supplied to the cells and a faster metabolic rate. Using the construction rule I just described, the way to make more capillaries is to build a bigger network with more branches. But this is not the network's only job. It must also fill three-dimensional space and have tubes that narrow and shrink so as to make the flow of fluid as smooth as possible and the journey to the capillary as short as possible. West, Brown, and Enquist found that this combination of criteria meant that a large body's network produced proportionately fewer capillaries than a small one. In a large animal each capillary must supply more cells. So each cell gets a smaller proportion of that capillary's supply and must slow down its energy consumption accordingly. An animal's metabolic rate is proportional to its number of capillaries.

If the number of capillaries was a simple linear function of body size every animal would have the same relative metabolic rate, directly proportional to body mass. But bigger animals have relatively fewer capillaries. In fact, the rules for the optimal network produce a number of capillaries that is proportional to body mass raised to the power of 3/4.

This, West, Brown, and Enquist believe, is why metabolic rate is proportional to the 3/4 power of body mass. It is the number, and therefore surface area, of capillaries that is similar to body size in the same way as metabolic rate, not the relationship between their external surface and volume, as Rubner thought, or the amount of muscle needed to stop them from buckling under their own weight, as McMahon suggested. In bigger animals, resources have a longer journey to the cells and each capillary must supply more cells, so the cells must slow their energy consumption. The network theory predicts that, as body size grows, cells must slow their metabolic rates proportional to body mass raised to the power of $-1/4$. A cell's metabolism goes as fast as it can, given the rate at which its body can provide it with resources. West compares cells to cars. A car might have a top speed of more than 120 mph, but it will not go nearly that fast in city traffic. Cells are similarly limited by all the other stuff around them. But if you take cells out of bodies and grow them in a dish, like a sports car on a country road, their metabolic rate speeds up until it hits maximum. Other scaling relationships, such as the breadth and length of the aorta, also fall out of the mathematics used to describe the ideal tubes. The theory predicts that they should be proportional to the 3/8 power and the 1/4 power of body mass, respectively. Measurements of different species show that this is indeed so.

West, Brown, and Enquist's model is an abstraction. It is an engineering solution derived from the laws of physics, rather than from any foreknowledge of biological systems. If you were going to design an animal, the model says, this is how you should do it. It is impressive that such a design reproduces key properties of living things, such as the way that blood vessels and metabolic rates change with body size. Real animals are bound to fall short of the theoretical ideal—their networks will not reproduce exactly the geometry of the model, nor will they change with size in the way it predicts. And in nature, metabolic rate is seldom if ever exactly proportional to the 3/4 power of body mass. But this shortcoming does not sink the whole theory. It would be very surprising to find a moose that ate the exact proportion of pondweed and land plants that, to the nearest gram, gave it the best diet. To see if moose eat optimally, we need to look at lots of them and see if their average diet is close to the theoretical prediction. Likewise, no animal is going to have a blood system built exactly according to the fractal rules. Network design is flexible—long-term exercise leads to more capillaries in a muscle, for example. And cells are flexible in the amount of energy they demand. But if there are underlying rules of metabolism, if we look at enough animals across a broad enough range of sizes, we should expect to see this theme underlying the variations that different species and individuals can play on it. And we do.

After doing some gnarly mathematics to explain how properties of the blood system, such as the jerky flow produced by a beating heart, would affect nutrient supply to the cells, West, Brown, and Enquist sent their model off to *Science*. Their paper was accepted, but only after several rounds of review involving eight reviewers (three or four is more usual). Three of the eight thought it was outstanding, revolutionary even. The Nobel Prize may have been mentioned. Three thought it was good. Two thought it was the dumbest thing they had ever seen. Lots of things about the fractal theory rub some biologists the wrong way. It is based on mathematics and physics, it is an abstraction, and it is a generalization. It throws away most of the nitty-gritty of biology and most of what most biologists spend their time doing. None of these characteristics are likely to recommend it to biologists who favor staring over squinting—the physiologist who has spent a career describing the quirks of the cardiovascular system or a biochemist who spends his or her time describing the reactions of cellular respiration. And not everyone believes in Kleiber's rule. Some biologists still think there is no one scaling law for metabolism and so nothing to explain. To them, West, Brown, and Enquist are squinting so hard that their eyes have closed.

It's fair to say, however, that D'Arcy Thompson would have been delighted with the network model. Among his papers is the following note, similar although not identical to the passage in *On Growth and Form* that discusses blood vessels:

> To keep up a circulation sufficient for the part and no more, Nature has not only varied the angle of the origin of the arteries to suit her purpose; she has regulated the dimensions of every branch and stem, and twig or capillary. The normal operation of the heart is perfection itself; we are told that even the amount of oxygen which enters and which leaves the capillaries is such that the work involved in its exchange and transport is a minimum. This perfect fitness, this maximal efficiency, we accept as a cardinal hypothesis; and we come to understand the form and dimensions of this structure or that by solving the problem of the work which they do.

That's what I was trying to say. Sometimes I wonder why I bother.

Networks in Plants

A tree looks like a naked fractal. In its branches can be seen the form of its distribution networks. But unlike blood vessels, the network of xylem tubes that carry water around the tree does not start with one big tube in the trunk and split into lots of narrow tubes as it approaches the leaves. Xylem is more like a bundle of electrical cables than plumbing; it is a group of tubes of the same size that split from each other as they near their destinations. Trees prompted Enquist to start thinking about networks, but could his theory apply to them?

In fact, plant networks reproduce many of the same features as animal networks, although for different reasons. As Leonardo saw, trees have area-preserving branching. And xylem vessels also taper, getting

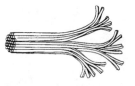

The water-carrying vessels of plants begin as a bundle and then branch to go their separate ways.
Credit: Reprinted with permission from *Science*, vol. 272, p. 122. Copyright 1997 by the American Association for the Advancement of Science.

narrower as they reach the treetops. The problem with a tube of a constant diameter is that the longer it gets, the more work it takes to force fluids down it. Resistance to flow increases down the tube. If xylem vessels kept the same breadth all along their length, the top leaves of a tree would never get any water. But if a tube tapers at just the right rate, the resistance to flow remains the same along its length. The degree of tapering that keeps the pressure in a tube uniform regardless of how long it is—and so keeps all parts of a plant equally supplied—plugs into the fractal model to give a metabolic rate predicted to be the 3/4 power of the plant's mass.

The mysterious number 4, the reason that scaling laws are built around multiples of one-quarter, and not the one-third powers that regular geometry predicts, is a consequence, as J. B. S. Haldane saw, of the pressure that natural selection exerts to fill an organism's volume with as much surface area as possible. In combination the networks for transporting resources and the surfaces for processing them become a four-dimensional entity. Three of these dimensions come from the two-dimensional area of surfaces, the capillary walls across which resources travel. They are folded so much, and fill space so well, that they take on the geometrical properties of a three-dimensional solid. It's as if you saw a ball, but realized on closer inspection that it was a crumpled sheet of paper. The fourth dimension is the distance resources must travel—the length of the tubes leading from point of uptake to point of delivery. Together, internal surface area and transport distance determine the rate at which an animal can burn energy. The number 3 in 3/4 arises because this four-dimensional network must fit itself inside a compact body existing in a three-dimensional world.

Alternative Routes

West, Brown, and Enquist do not have the field to themselves. While some biologists were accusing them of oversimplifying biology in trying to explain metabolic rate, some physicists thought their model wasn't simple enough. "We read the paper, and it seemed somewhat complicated," says Jayanth Banavar, a physicist at Pennsylvania State University. "We thought there must be some other explanation."

Banavar teamed up with two other physicists, Amos Maritan and Andrea Rinaldo, and a biologist, John Damuth, to design a network using different criteria. Rather than looking for a network that maximized energy supply, they sought one that minimized the rate of flow in the system and so the volume of blood. They considered a body as a set of delivery stations, cells, serviced by a network of pipes stemming from a single source. Unlike the fractal model, the network can flow through its destination and go on to somewhere else. Cells in this model are not like twigs at the end of a branch. They are more like stations on a railway line, where some people (resources) get off. As the network expands, the amount of fluid in the system must rise. The team showed that, in the most efficient three-dimensional network, where the distance from source to destination is minimized, the amount of fluid must rise proportional to the volume of the network raised to the fourth power.

This model got the desired number 4—and, incidentally, also predicted the geometry and flow rates of river networks—but it created a problem. As a network, or an animal, grows, the amount of fluid that is in transit at any time, and so not available to be used, must increase. So for the rate of supply to keep pace with network size, more and more fluid is needed. If animals were like this, big ones would need a volume of blood larger than the volume of their body. So Banavar's team thought again and forced their model network to stay contained inside the body. In this case, as the network grows, its capacity to supply resources declines, and so the cells receiving the resources must adjust their demands accordingly. The network that minimizes the amount of blood in transit delivers resources proportional to the 3/4 power of the body volume it supplies—as predicted by Kleiber's rule.

For Banavar it is the fact that networks are directed—the flow only goes one way—rather than that the network is a fractal that is the key point. "The question," he says, "is always 'What are the essential features and what are less important?'"—which details you need, and which you can junk. The fractal model may be accurate, says Banavar, but he sees it as a special case of a more general set of networks described by his team's more economical model. Unlike West, Brown, and Enquist's, this model does not need the complications of fractals, area-preserving branching, or fluid dynamics. Nor does it assume that the networks get resources to the cells as quickly and efficiently as possible. Instead, Kleiber's rule emerges as a general property of distribution networks—not just the best way to deliver resources but the only way. West, Brown, and Enquist point out that their model is better at predicting specific details about living things—mirroring some of the criticisms aimed at them by biologists who believe that the fractal model is too general and abstract.

The original West, Brown, and Enquist model reactivated interest in metabolic rate, and since they launched their theory still others have joined in with ideas and theories quite different from the two network models. One recent model based on the chemical structure of cell membranes argues that small animals' membranes are softer and leakier than those of large animals, and so they use energy and process chemicals more quickly. Another argues that metabolic rate arises from the way that the amount of DNA in a cell affects its size (the more DNA an organism has, the bigger its cells). Still another looks at metabolic rate from the viewpoint of how body size affects an animal's ability to obtain food and store it in the body. And another argues that no one factor depends on metabolic rate but rather that Kleiber's rule is a consequence of the "allometric cascade" of the many different processes, involving the heart, lungs, and cells, that make up metabolism. And there are others I have not mentioned.

But in this crowded field the West-Brown-Enquist model is currently the front-runner. Some reasons for its lead are scientific. The network model's mathematics are complex, but its foundations are intuitive and its applications universal—all organisms must get stuff from somewhere to somewhere else. It's difficult to see how else organisms could be constructed. The fractal model also makes predictions that match other aspects of biology, particularly its calculations of the width of tree trunks and aortas. The weakness of the network models is that they are difficult to test—it's hard to see a killer measurement that might settle things one way or the other, particularly as these models are generalized abstractions of how life works. Other reasons for its preeminence are political: The team got there first, they have a long list of publications in the most prestigious scientific journals, and, as we shall see, they have been vigorous in using their model to explain other aspects of biology, which has raised their profile.

Nine years, at the time of writing, after the fractal model's debut, researchers are divided along similar lines, in similar proportions, as the original reviewers that *Science* got to comment on the theory. Some are keen advocates, some are virulent opponents, and the majority are interested but undecided, waiting to see which way things go. The debate is still fierce, but no one has found a fatal flaw in the fractal theory to convince the scientific community that it is invalid. "If it's wrong, it's wrong in some really subtle way," says West.

Single Cells, Virtual Networks

Another branch of the tree of life challenges the network theory. Many, probably most, species don't bother sticking their cells together into organs, networks, and bodies. They live as single cells. But they are not beyond the reach of biologists looking for scaling laws. Single-celled organisms certainly come in lots of different sizes. The alga *Ostreococcus tauri*, discovered in the Mediterranean in 1994, has cells a millionth of a meter across. Another alga, *Acetabularia*, also known as the mermaid's wineglass, has cells that are 5 centimeters long and 1 centimeter across. By my reckoning this is almost 10 billion times the volume of *Ostreococcus*; a whale is a mere 10 million times more massive than a mouse. Other unicellular organisms are even larger, although they pull some fancy tricks in the process. The alga *Caulerpa taxifolia* grows thin fronds up to a meter long, each of which is basically a single cell, although one that is subdivided and reinforced. And some slime molds come together to form blobs that are tens of centimeters across and have no cell walls, although they may contain millions of cell nuclei (the bit of the cell containing the chromosomes). None of them have any blood vessels, xylem, or any other sort of plumbing. Yet there is evidence that the metabolic rates of unicellular organisms also scale as the 3/4 power of their mass. How can a theory based around distribution networks say anything about them?

In fact, unicellular organisms are not just blobs. Their resource distribution problems are similar to those of larger organisms, and they have come up with similar solutions. The cells belonging to the kingdom called eukaryotes, which includes plants, animals, fungi, and single-celled organisms such as algae and amoebas, have complex structures that are like miniature organs (the technical term is organelles), such as the fuel-burning mitochondria and plants' carbohydrate-making chloroplast. Like capillaries in the blood system, these structures provide a surface for the transactions of biochemistry to take place. And like capillaries, these miniature organs keep the same size in different-sized cells—they are the terminal units in the cellular network. Evolution should drive a cell to speed its internal transport and maximize the surfaces where molecules are used—in the same way that larger bodies increase the surface area of their lungs and intestines—so that a cell can get on with its life as quickly as possible.

Even in unicellular organisms, the evolutionary drive to use resources as quickly and efficiently as possible should produce some sort of structured distribution network. Such a network could be physical—cells have microscopic cables and packages that shunt molecules around—or it could be virtual, with resources flowing by diffusion from where they are taken up at the cell membrane to the terminal organelles inside the cell. The distance from where cells take up chemicals to where they use them is analogous to the length of the tubes in the blood or xylem system, and as in larger organisms, evolution should seek to minimize it. This means that cells' metabolic rates should scale as the 3/4 power of their body masses, regardless of the considerations that apply to tubular networks, such as how they branch or how fluids flow through them.

The metabolic rate of an isolated mitochondrion, sitting in a lab dish, also falls on Kleiber's line. Even organelles have their own distribution networks. At the molecular level, their respiration is run by large, complex molecules. And even in these isolated molecules, the reactions of respiration run at the speed predicted by Kleiber's rule, suggesting that within these molecules' structure there are nanoscale distribution networks that move individual oxygen and ATP molecules around. From monsters to molecules, this line of reasoning extends the network theory's reach over a size range of 27 orders of magnitude, or a thousand trillion trillion times. The same logic extends the theory to the bacteria, the vast group of single-celled organisms lacking mitochondria, chloroplasts, or nuclei, but that do have these large and complex molecules. West, Brown, and Enquist's model describes these as "virtual fractals": Resources are distributed in a branching pattern through the cell, even if they are not in tubes. For the model devised by Banavar and his colleagues, this isn't such an issue—resources spread out from a central point, like food being shared from the head of a table to the guests seated around it.

It seems that natural selection has such a strong preference for efficient distribution that the same fractal network solution has evolved many times, at scales from molecules, to cells, to plants and animals, taking on many different forms and for many different substances but always converging on the same fundamental properties. These networks are such a versatile solution to the problem of supplying a body with resources that they have allowed life to evolve into a remarkable range of sizes. It's as if a human engineer had invented a single mechanism that could power everything from silicon chips to supertankers.

So model networks can explain why the slope of the line linking mass and metabolism has a gradient of 3/4. But that leaves lots of variation. Animals of the same size in different groups of organisms have very different metabolic rates. Reptiles are slow, birds fast, and mammals somewhere in between. Plants are slowest of all. Even accounting for size, the speediest metabolic rates seen in nature run 200 times more quickly than the slowest. Body size controls the rate at which cells can be supplied with resources, but this is not the only form of energy that influences metabolic rate. There is also the rate of the chemical reactions within cells. And this depends on temperature.

Of Reptiles and Bananas

I met Jamie Gillooly in an Albuquerque coffee shop. He picked up a banana skin. "If you correct for size and temperature," he said, "this banana on the table, at least before I ate it, was respiring at a rate similar to everyone else in this room."

In autumn 1999, Gillooly was completing his Ph.D. at the University of Wisconsin on the effects of size and temperature on the plankton in lakes. Biologists who study aquatic systems do not pay much attention to the work of those concentrating on land life, so he had not come across the scaling theory until Jim Brown came to Wisconsin to give a talk on it. The two met over breakfast to discuss their ideas; by the time they finished, they were "both foaming at the mouth with excitement," says Gillooly. A few months later Gillooly was on his way to New Mexico to do a postdoc.

Life runs faster in the warm—this is why fridges slow the growth of mold. Temperature has an exponential effect on metabolic rate. A 5°C rise in body temperature will raise metabolic rate by 150 percent. It was West who suggested how to incorporate this effect into the equation for metabolic rate. The answer, he said, was the Boltzmann factor—named after the German physicist who believed that life was a struggle for energy and who also laid the foundations for statistical mechanics, the physical theory that explains how particles behave en masse. The Boltzmann factor is the probability that two molecules bumping into each other will spark a chemical reaction. The higher the temperature, the faster the molecules move, the harder they collide, the greater the probability of a reaction, and so the faster the chemical process. Like metabolism, the effect is exponential. West, Gillooly, and the rest of the team added the Boltzmann factor to Kleiber's allometry equation relating metabolic rate to body mass raised to the power of 3/4.

The results were dramatic. Accounting for temperature in this way mopped up much of the variation in metabolic rate that size alone could not explain. Correcting for temperature also brought different groups onto the same line: Reptiles' relatively slower metabolic rates are a consequence of their lower body temperatures, showing that cold- and warm-blooded animals share fundamental metabolic processes. The same goes for plants and animals.

Like the network model, this is another simplification. The chemistry of metabolism involves many reactions, each of which will require a certain amount of energy to get going. Applied to metabolism, the Boltzmann factor is a black box. It could be a kind of average for all the hundreds of chemical reactions in metabolism, or it might be the energy needed to get over one crucial hump in the path. But including it reduces the size-adjusted variation in metabolic rates from a factor of 200 to a factor of 20.

Perhaps this remaining 20-fold variability is some indication of the wiggle room that organisms have that enables them to tinker with their metabolic rate to match their circumstances. Animals living in cold environments might crank their metabolic rates above the grand average, whereas plants in impoverished soils might depress theirs below it. Or perhaps it is a measure of how far organisms can deviate from the optimal network before they become so inefficient that natural selection weeds them out. This leftover variation, if you like, is what really needs to be explained—the place where biology will come into its own.

"I'd like for biology to have a sense of average idealized organisms that share similar principles, which can be understood mathematically—and that what you should be studying are the deviations from that," says West. "At the end of this century there'll be a fantastic theory that'll predict lots of things—everything from ecology to how genes are turned on and off. And I'd like it if there were a footnote in the textbooks saying: 'At the end of the twentieth century people started taking the problem seriously, and there were these guys who pointed the way to getting rid of the uninteresting part of the problem, the bit that depends on mass and temperature, that determines 90 percent of what we see. Now this *real* theory of biology is devoted to the other 10 percent.'"

6   THE PACE OF LIFE

Every living thing grows, changes, and, if it is among the lucky few, reproduces. Then, if it doesn't first get eaten, freeze, drown, starve, contract a lethal infection, fall out of a tree, or encounter any of the other fatal hazards of life in the wild, it begins to fall apart. And then it dies. Listen closely to the hum of metabolic rate, and you will hear this rising and falling melody of life.

Any organism that just kept ticking at its resting metabolic rate wouldn't get very far. All must obtain a surplus of energy and then invest it wisely. Natural selection keeps score of how well organisms exchange one currency of life, energy, for another, offspring. Successful investment strategies multiply; bad ones go out of business.

Evolution has smiled on an extraordinarily diverse portfolio. Consider animals. They are cold- and warm-blooded, herbivorous and carnivorous. Some start life as eggs in a nest and lay eggs themselves. Others gestate in their mother's womb and bear live young in their turn. Some grow strong on their mother's milk. Some are fed by their parents until they are adults. Others are shot into the plankton as unfertilized eggs and must fend for themselves from that point onward. Some are recognizable in youth as miniature versions of their adult selves. Others go in for a complete remodeling before adulthood, breaking down their juvenile bodies and using the parts to make something very different. Some live for more than a century. The water flea *Daphnia* completes its life cycle inside a week.

It's easy to imagine what an all-conquering evolutionary super-beast—a Darwinian demon—would be like. It would grow to large size and sexual maturity in a flash and produce vast numbers of hearty offspring, forever. Such a paragon would monopolize the energy available to life. But, of course, this isn't possible. Life's resources are finite. Every feature of every plant and animal represents a choice made by evolution about how to invest energy. And every choice precludes other choices. Organisms must choose whether to put their resources into muscular legs or wings, gaudy tails or flowers, fearsome horns or juicy fruit. They must also choose what to invest in and when: whether to become sexually mature at a young age, and get on with mating and breeding before any mishaps befall them, or delay reproduction until they are bigger and better equipped to compete for resources or mates. They must choose whether to produce a few large offspring, each of which stands a good chance of surviving, or lots of little ones, most of which will die young. When they reproduce, they must decide whether to hold something back for future breeding seasons or fling everything on the one shot and die in the attempt. The result of all these decisions is called an organism's life history.

Choice—trade-offs, in other words—is fundamental to life's design. And there are still deeper unities to be pursued. Even in a finite world, the diversity of life makes it look as if there are an astronomical number of biological options. In fact, metabolism controls what choices can be made, and understanding metabolic rate peels back the multiplicity of life histories to reveal one sleekly efficient mechanism for turning energy into offspring.

The Form of Growth

D'Arcy Thompson noticed that organisms grow quickly in their youth and then ever more slowly thereafter. Some, such as birds, mammals, and insects, stop growing when they reach adulthood and start breeding. Others, such as fish and molluscs, seem to keep gaining weight throughout their lives, but this rate nevertheless decreases as they age. For all sorts of animals, plotting their size against their age creates an s-shaped curve, which starts shallow, takes a rapid turn upward, and then levels off. An animal's growth rate seems to be some function of its age. Thompson keenly sought the laws of growth that controlled this trajectory, for he believed they controlled animal form. But in *On Growth and Form*, he had to admit defeat:

> For the main features which appear to be common to all curves of growth we may hope to have, some day, a simple explanation. . . . The characteristic form of the curve of growth . . . is a phenomenon which we are at present little able to explain, but which presents us with a definite and attractive problem for future solution.

By the time the second edition came out in 1942, someone had offered that simple explanation.

Born near Vienna in 1901, Ludwig von Bertalanffy got his Ph.D. at that city's university in 1926, for a study of the work of Gustav Fechner, a nineteenth-century German psychologist and philosopher who studied human vision. Soon after, Bertalanffy concluded that the life sciences had reached an impasse. Knowledge was being churned out faster than ever before, but without theory the babble of unconnected facts drowned out understanding. "Today biology is in its pre-Copernican period. We possess an enormous mass of facts, but we still have only a very incomplete insight into the laws governing them," he wrote. "Only if the multiplicity of facts is ordered, brought into a system, subordinated to great laws and principles, only then does the heap of data become a science. . . . The chief task of biology must be to discover the laws of biological systems to which the ingredient parts and processes are subordinate. *We regard this as the fundamental problem for modern biology.*"

Bertalanffy saw the impasse as an opportunity as well as a crisis. He took inspiration from physics, then convulsing as quantum theory replaced the Newtonian worldview. Perhaps an equally radical shift in biological thought could reap similar rewards. He began to ponder the fundamental properties of life. The two he settled on are themes we have already encountered repeatedly: hierarchical organization and the continual exchange of matter and energy with the environment. Just as quantum theory turned subatomic particles from rigid, fixed points into waves, so Bertalanffy sought to replace the idea of organisms as static, self-contained entities with something more dynamic and fluid. "Living forms are not *in being*, they are *happening*," he wrote. "They are the expression of a perpetual stream of matter and energy which passes the organism and at the same time constitutes it."

Hierarchical organization meant that one could not hope to understand organisms by considering any aspect of their biology in isolation. The processes and structures of life were inseparable from, and only made sense in the context of, the organisms in which they were housed. Bertalanffy called this research program "organismic biology." Through the 1930s and 1940s—following Alfred Lotka's lead—he expanded the scope of his program beyond biology, coming to believe that all systems made up of many interacting elements shared similar properties, regardless of whether the system's elements were physical, biological, psychological, or social. The task, Bertalanffy argued, was to work out the common principles that controlled these systems. In the process, biology and physics would unite. But instead of explaining the former in terms of the latter, he imagined a new science, based on logic and mathematics, that would provide a higher level of explanation and of which all other sciences would be a subset. It would be called General Systems Theory.

These ideas should sound similar to what is done at the Santa Fe Institute, and indeed Bertalanffy helped found what has become the study of complex systems and emergent properties. But while he pursued such ambitious syntheses, Bertalanffy also carried on lab work, studying growth and metabolism. In his experimental work, just as in his theoretical pursuits, Bertalanffy sought to find the general principles amid the mass of detail. Growth involves myriad constantly varying chemical reactions, and every individual of every species will be different at every moment. But, he reasoned, just as we regard a company's statement of profit or loss as a meaningful generalization, even though it conceals a host of individual transactions, so it ought to be possible to make generalizations about animal growth.

In 1934, Bertalanffy suggested that an animal's growth rate was proportional to its metabolic rate. He reasoned that the speed of an organism's growth must depend on the speed at which it gets and uses resources from the outside world. Bertalanffy thought there were several different metabolic types, some matching the surface law, others not, that created several different growth trajectories. Since then other measurements have shown that—with the usual caveats about variation—the range of metabolic types that Bertalanffy saw was an illusion and that growth rates, like metabolic rates, are in general proportional to body mass raised to the power of 3/4. Small animals grow relatively quickly, big ones relatively slowly.

Bertalanffy believed that the link between metabolism and metabolic rate could explain why growth took the form it did and why it eventually stopped. He suggested that the net rate of growth was a simple balance between income and expenditure, reflecting the difference between the energy needed to build and expand bodies and the energy needed to maintain, repair, and replace existing parts. Because big animals metabolize relatively slowly, their maximum growth rate is also slower than that of small animals. But, he argued, this isn't true for the costs of upkeep: The amount of energy needed to keep what you've already got in working order grows linearly with body mass.

So growth becomes a function of size, rather than time. Growth slows with age because as an animal grows, it comes to spend all its resources on maintenance, leaving nothing for further expansion. Bertalanffy came up with a simple equation expressing this balance between growth and decay, which fits the s-shaped curve very nicely. It has since been used in fisheries and forestry to calculate the best stage at which to catch fish or cut down trees. Roughly, this is the point at which growth has slowed so much that it is no longer economical to leave a pine in the ground or a cod in the sea, but rather is more profitable to harvest it and start over with a smaller, and so faster growing, plant or animal. If a tree is adding wood, or a fish flesh, at high speed, it is becoming more valuable, and it pays to leave it be. If it is growing only slowly, it pays to asset strip, and invest the money in something more profitable. This is why industries based around harvesting slow-growing animals, such as whales or old-growth forest, or hunting large primates for meat, are hard, if not impossible, to sustain—from a coldly economic viewpoint, the best strategy would probably be to slaughter every whale and buy Google stock, or something, with the proceeds.

But without an explanation for why metabolic rate slows with size, this equation, although useful, says nothing about the biology of growth—it just fits the curve. Why should big animals find it harder to make a metabolic profit than small ones? They might need more food, but they are also better equipped to get it: They can cover more ground more quickly than small animals, tackle bigger food items, and, if they are herbivores, digest it more fully. And their slower metabolisms are also presumably cheaper to maintain.

The network theory offers a new perspective on growth. It builds on Bertalanffy's reasoning to show why increasing size eventually halts an organism's growth. The difference compared with Bertalanffy's theory is that where he focused on demand, West, Brown, and Enquist are more concerned with supply. Their theory predicts that the amount of tissue needing to be fed rises more quickly than the network's power to deliver food. As an animal gets bigger, each of its capillaries must supply more cells. Eventually, the network can supply only enough energy for maintenance, and growth must stop. The bigger a body gets, the harder it is for its networks to supply the distant outposts of its body, and, like an overstretched empire with failing supply lines, further expansion becomes impossible.

What this means is that, as an animal grows, the proportion of its resources that it puts into growth must decline. Proportionately, the growth trajectory for all animals is remarkably constant. An animal that is one-fifteenth of its final weight, whether it is a 40-kilogram calf or a 1.5-kilogram cod, will invest about one-half of its resources in growth. Once it reaches half its final weight, this proportion has dropped to about 15 percent. Life span and the amount of growth, of course, vary hugely for different species: A 40-kilogram calf is a newborn, whereas a cod takes six years to reach 1.5 kilograms and will probably never reach its upper size limit of about 25 kilograms.

What goes for animals also goes for plants. To pursue his botanical interests, after he completed his Ph.D. at New Mexico, Enquist began collaborating with plant biologist Karl Niklas of Cornell University. Niklas's speciality is plant biomechanics—he tries to work out why trees don't fall down. Studying this question has made him into one of a relatively small group of plant biologists expert in allometry. Niklas was and is unsure about whether the fractal theory really explains the scaling of metabolic rate; he has written that understanding scaling is "as potentially important to the biological sciences as Newton's work was to the physical sciences," but he is one of those who sees the fractal model as untestable. Nevertheless, he was impressed by its generality. "I've gone from being an atheist to an agnostic" is how he describes his view. And whatever the mechanism, Niklas was convinced that scaling laws were powerful and general descriptions of plant biology, that the same laws applied to all plants from pond algae to forest giants, and that metabolic rate lay at their heart. He was well qualified to bring a new source of mathematical oomph to complement Enquist's ecological expertise.

It was time for another trip to the library and another bout of biblio-ecology. Enquist and Niklas compiled data on the growth rates of plants from unicellular algae, which you can watch dividing under the microscope, to the giant sequoia in California known as General Sherman—the most massive tree in the world, which you have to visit each year with a very long tape measure. The General is 20 orders of magnitude, or 100 million trillion times, larger than the alga. But regardless of their evolutionary ancestry or habitat, Enquist and Niklas found that all plants' growth is proportional to the 3/4 power of their body mass. That is, all plants share the same basic strategy for turning solar energy into vegetation. And animals and plants show similar growth patterns: Trees grow at the same rate as animals of the same size.

No central controller sets the growth rate. Growth emerges from the network, from the interaction between a body's demand for energy, its network's capacity to supply the body with energy, and the balance, set by evolutionary experience, between growth and other demands on energy. In a healthy body, cells do not grow at the expense of their neighbors, because they share genes—and so evolutionary interests. But sometimes a genetic mutation causes part of the body to rebel against its neighbors, and against the limitations of the network that serves it. Such cells increase their metabolic rate and stimulate new blood vessels to grow to supply them with resources. Such cells that do not know when to stop growing form tumors and cause cancer.

In 2003 a team of Italian and American researchers showed that tumor growth also follows the growth law described by West, Brown, and Enquist, implying that a tumor's nutrient supply limits its growth. It was already known that any cancer cell had to be within 0.1 millimeters of a blood vessel to divide. The team hopes that relating a tumor's size to its growth rate will permit calculation of the drug dosage needed to reverse a tumor's growth and also to predict when malignant tumors will metastasize and spread around the body. Cancer is a diverse disease, however—every tumor is different in its genetic makeup and biological effects, and it is still unclear how discovering such general principles might help treat individual clinical cases.

Changes of Life

Getting bigger is only one part of life's history. Many things besides size change as an organism ages. It moves from egg, or womb, into the world. It leaves its mother's care and tries to feed itself. It becomes sexually mature. It may go from a larval stage into an adult body. All these changes in lifestyle have their attendant changes in biology and their own set of choices. The time that organisms spend in the egg or the womb, the length of time they depend on their parents, and the interval between birth and adulthood all vary enormously. But relatively, these times have much in common. The average mammal, for example, spends about 4 percent of its life in the womb—which means three weeks for a rat, 50 days for a fox, five months for a goat, and eight months for a hippo—and reaches sexual maturity a fifth of the way through its life.

The duration of all these phases—the time spent in the womb or egg, the period between birth and weaning, and the time taken to reach sexual maturity—is proportional to body mass raised to the power of 1/4, the power law that describes biological times. Like growth, the course of an organism's development is controlled more by its size than its age, and considering size reveals the unity between species. Correcting for size accounts for much of the variation in developmental rates, and correcting for body temperature accounts for some of what's left over. Together, the two variables explain more than three-quarters of the variation in hatching times for fish, insect, amphibian, and bird eggs. If a salmon egg could somehow be expanded to the same size as a hen's egg, and then incubated at the same temperature, they would hatch after approximately the same length of time. The two would also take approximately the same amount of time to reach maturity.

There is more to growth and development than energy. A builder with a generator but no bricks would build nothing. Likewise, organisms need materials as well as calories: A body won't work properly without a balanced diet and what it can do depends on what chemicals it gets. Just as they vary in size and temperature, organisms also vary in their chemical composition, and this variation can explain another tranche of their biological differences.

After carbon the two most important chemical elements found in living things are nitrogen, which is a large component of proteins, and phosphorus, which is found in ATP, the energy molecule, and also in the molecules that carry genetic information, DNA, and its close cousin RNA. In many organisms, such as plants, it seems to be the availability of these elements, rather than solar energy, that limits growth. It is these elements that chemical and organic fertilizers add to soils. Bone meal, for example, is rich in phosphorus.

Different species contain similar proportions of protein and so nitrogen. The amount of phosphorus is much more variable. Fast-growing species have more phosphorus because RNA is part of the cell's protein-making machinery, so fast-growing cells need more of it. RNA is also used in the process that turns the information in the genome's DNA into protein, giving hard-working cells an extra demand for phosphorus. Animals with high growth rates therefore also contain a lot of phosphorus. And when the metabolic ecologists compared relative phosphorous content to growth rate, they found that this quantity explains some of the variation in developmental times not accounted for by body size and temperature.

Chemistry, then, is a third factor, besides body size and temperature, that controls metabolic rate. To explain metabolic rate and how it sets life's other rates, we must consider materials as well as energy and organisms' chemical composition as well as their size and temperature. Life both depends on and controls a supply of elements through the rate at which plants take up nutrients and the rate at which they pass up the food chain and ultimately return to the world via death and decomposition. This chain links the events in cells to processes that encompass the whole earth. In 1934, Alfred Redfield noted that the ratio of carbon to nitrogen to phosphorus in seawater around the world was the same as the average ratio of those elements in the bodies of marine plankton. There are 106 carbon atoms, to every 16 nitrogen atoms, to every 1 phosphorous atom. The dietary habits of unicellular plants, therefore, control the chemistry of the world's oceans—although no one has been able to explain why the Redfield ratios should take the values they do. As Lotka predicted, chemistry, like temperature, is another controller of biology that humans are changing on a global scale. Since 1908, when the German chemist Fritz Haber worked out how to take nitrogen out of the air and turn it into ammonia, the quantity of nitrogen available to living things has doubled. Much of this nitrogen is washed into lakes and rivers—U.S. rivers contain between two and 20 times more nitrogen than they did before the Industrial Revolution. From rivers the nitrogen goes into estuaries and coastal waters, where it stimulates plant growth, sometimes leading to algal blooms. When these blooms decay, the bacteria that eat them use up oxygen, sometimes choking animal life. Each summer, several thousand square kilometers of the Gulf of Mexico becomes a low-oxygen "dead zone," thanks mainly to agricultural runoff from the Mississippi River.

Evolution's Winning Post

All growth and development are just means to the end of reproduction. It doesn't matter how big you are, or how quickly you grow, if you don't leave any descendants. Sometimes reproduction and growth are indistinguishable. When a single-celled organism such as a bacterium divides, a new individual is born. For multicellular animals reproduction is also growth by proxy, although in a more complicated fashion. Instead of investing its energy in itself, an animal puts its resources into a copy of itself (a partial copy, if reproduction is sexual). This process presents a further array of choices: how much to invest in the offspring; whether to divide this investment between a few big ones or lots of small ones; and after birth, how long to go on feeding and protecting your progeny.

Metabolism sets the boundaries of reproductive choices, too. The rate at which animals can seize energy for themselves limits the rate at which they can pass it on to their offspring. So, not surprisingly, many aspects of reproduction march in step with metabolism, and an animal's size is the best guide to its reproductive biology. Bigger animals produce bigger young, but they devote a smaller proportion of their resources to their offspring. The clutch of a 100-kilogram ostrich consists of thirteen 1-kilogram eggs, totaling about one-eighth of the mother's body weight. A 3-gram hummingbird can lay two eggs that each weigh a quarter as much as itself. Big fish and reptiles also lay proportionately lighter clutches than small ones. The mass of mammals' litters shows a similar trend, and big mammals' milk contains less protein than that of small species. All these properties are more or less proportional to body mass raised to the power of 3/4, although there is much variation about this number. One thing that does vary between groups is the way these resources are divided up: Birds and mammals produce relatively few large offspring, in which they invest considerable resources, both before and after birth. Many reptiles, fish, and invertebrates go for quantity rather than quality, producing many small young, each of which stands only a small chance of survival. Bigger animals also breed less often—the interval between litters or clutches is proportional to body mass raised to the power of 1/4—and so produce a smaller number of offspring over their lifetime.

The 30-Tonne Gorilla

The bigger you are, the more energy you can get hold of and the more offspring you can produce. One species, however, flies in the face of this apparent evolutionary no-brainer. An adult female gorilla weighs about 100 kilograms and will give birth to between three and six young in her life. The average European or North American woman weighs considerably less but will produce approximately two babies.

Countries' birth rates fall as their economies grow. Less developed nations typically have both high birth and death rates. As health improves, the death rate falls, and population rises quickly. But eventually birth rate falls too, and the population stabilizes. In parts of Western Europe, the birth rate has fallen so much that the population is now in decline. It seems odd that the richer people become, the more slowly they breed. One could say that people don't need large families in urban societies with low death rates, but in evolutionary terms, reproduction has nothing to do with need: It's about making as many copies of your genes as possible. Humans are not blind slaves to their genes, and society and technology can influence reproductive decisions in many ways. But it still seems curious that customs and fashions in rich countries run so strongly against large families.

In fact, humans reproduce exactly as much as their metabolisms predict. It's just that, unlike every other species, biological metabolism makes up only a small fraction of human energy consumption. We also use energy in our heating, air conditioning, cars, appliances, and so on. Our societies use it in manufacturing, airplanes, communications, and road building. Jim Brown and his student Melanie Moses compared per capita energy consumption and birth rates for more than 100 countries. They found that birth rate falls smoothly as energy consumption rises. The average U.S. resident burns energy at a rate of 11,000 watts, about 100 times the human metabolic rate. Raising a child with similar demands, and equipping him or her with the advantages and status necessary to compete in the world, is hugely expensive. Perhaps this explains why energy-hungry families produce so few offspring. According to figures released in April 2005 by the Liverpool Victoria Friendly Society, it costs £152,000 (about US$250,000) to raise a British child from birth to age 21, rising to £300,000 if you have him or her educated privately.

The relationship between energy consumption and fertility of humans in different societies and for different periods in history follows the same pattern as that seen for other primate species. The human reproductive rate is that of a great ape with outsized energy demands. A woman in the United States uses as much energy as a 30,000-kilogram primate would, if such a behemoth existed, and she produces the same number of children. Picture a billion or so 30-tonne gorillas stomping across the earth, and you get a good idea of why humans have such a profound effect on their environment.

It is a mystery how humans' extra-metabolic energy use might influence their breeding decisions. Brown and Moses speculate that each society is an energy distribution network with people rather than capillaries as its terminal units. As a nation's ability to use energy increases, so does the size and cost of the infrastructure needed to supply that energy. Like a cell in a large animal, people plugged into the largest energy supply networks might only be able to reproduce slowly.

To succeed in life, organisms must get hold of as much energy as they can and use it in the best way. West, Brown, and Enquist's theory charts the boundaries of what is possible—the maximum energy that an organism can use—from physics and chemistry. The team's subsequent studies of metabolism, growth, and development seem to show that everyone still in the game is pushed up hard against these physical boundaries, using energy as fast as they can, thanks to the fractal networks that transport resources around their bodies. Such networks have evolved many times independently in animals and plants. It is still not clear how strict the physical limits on biological systems are, but looked at over a large size range, there is a striking uniformity in how animals channel their energy into growth, development, and reproduction.

Burning Out

Metabolism sets the pace, and balances the scales, for the beginning of life, the course of life, and in reproduction for the purpose of life, at least in evolutionary terms. What about the end of life? That, after all, is what got Geoff West interested in metabolism in the first place.

In the inanimate world, putting a lot of energy through a thing is a good way to wear it out. The harder you work a machine, the quicker it falls apart. We talk about running things into the ground; people who work and play hard say they are burning the candle at both ends, and the candle that burns twice as brightly burns half as long.

First appearances suggest that the same is true for living organisms. Large animals with relatively slow metabolisms live longer. The match is striking: metabolic rate per cell declines proportional to body mass to the power of −1/4, and life span increases at a rate proportional to body mass raised to the power of 1/4. This means that every cell, be it rat or rhino, burns approximately the same amount of energy in its lifetime. Heart rate also declines in line with relative metabolic rate, as the −1/4 power of body mass. At rest a mouse's heart beats more than 500 times a minute. A shrew's blood races out of its heart, around its body, and back to base in four seconds. Our blood takes nearly a minute to complete the same journey. An elephant's heart beats about 30 times every minute. Other organs, such as the kidneys, also work much harder in smaller animals. Life spans vary proportionately. A mouse, even a pampered one in the lab, is unlikely to see three summers; an elephant can expect three score.

This means that every mammal should get about the same number of heartbeats—about 1 billion. One recent, unserious, calculation put the number at 955,787,040. Likewise, each mammal will draw about 200 million breaths before it expires. This has become folk wisdom. The astronaut Neil Armstrong once remarked that he did not jog because he believed that God had given him a finite number of heartbeats and he would be damned if he was going to fritter them away running up and down the street. It almost looks as if every animal— every cell—has a certain amount of energy to burn, after which it dies.

For many years, people thought just that. The notion that the rate of living determined the length of life was given scientific form by Max Rubner, he of the surface law, in 1908. Rubner used his calorimetry experiments to calculate the relationship between energy use and life span in horses, cows, cats, dogs, and guinea pigs. He found that each gram of guinea pig flesh burns 260 calories in its six-year lifetime and each gram of horse flesh burns 170 calories during 30 years of life. We now know that horses can live up to 50 years, which works out to 280 calories per gram per lifetime, even closer to the guinea pig. Other species fell somewhere in between. This fit well with Rubner's view that biology was a matter of food and energy. With characteristic grandeur, he announced that the finding had "the unity of a great law."

Experiments following on from Rubner seemed to confirm that, if you slowed an animal's metabolism, you extended its life. Cooler temperatures had the same effect as increasing body size: Fruit flies reared at 30°C live on average 14 days, those at 10°C live for 120 days. Water fleas show a similar link between temperature and life span. And because water fleas are transparent, you can see their hearts beating inside their bodies, showing that both cool and warm fleas have the same number of heartbeats in their lives. Rats that led sedentary lives lived longer than those that were forced to exercise. All this evidence made a deep impression on the man who would become most associated with the idea that to live fast is to die young—the American biologist and statistician Raymond Pearl.

Well over 6-feet-tall, loud and boisterous, funny and caustic, given to marathon bouts of French horn playing, and with apparently bottomless supplies of enthusiasm and self-belief, Raymond Pearl was perhaps the only swashbuckling statistician who has ever lived. His great friend was the journalist Henry (H. L.) Mencken, and during Prohibition he was part of the Saturday Night Club that met in Mencken's cellar to booze it up and play music. Pearl was an enemy of fundamentalism and Puritanism and caused a minor scandal in Prohibition-era America when he argued that health records showed that moderate drinkers outlived teetotallers. When he died in 1940 at age 61, Mencken wrote his obituary.

Pearl began his scientific career at the University of Michigan, earning a doctorate in biology for his work on flatworm behavior in 1902. He discovered the joy of statistics while working in London with the eminent statistician Karl Pearson in 1905. Pearson believed in laws of nature but of a rather different sort than those sought by Bertalanffy and D'Arcy Thompson. Pearson saw science as the classification of facts, and his scientific laws were the mathematical expressions that described the patterns in messy data. He saw no need to work out what might be causing those patterns. Such statistical techniques had barely reached America; Pearl resolved to introduce them. Statistical analysis, he argued, was not a piece of drudgery useful only for bringing recalcitrant experimental data into line—it was the foundation of biology. "The matters with which biostatistics concerns itself constitute some of the most fundamental and important problems of pure biology," he wrote. "Biostatistics . . . is the sign, the symbol, and indeed in some respects the very essence of the *biology of groups*."

By the mid-1920s, Pearl's charisma and newfangled skills had made him something of an academic celebrity—the highest-paid professor at Johns Hopkins University in Baltimore, free to study whatever he liked, and (many a researcher's dream) relieved of all teaching duties. Besides churning out a huge number of academic papers and more than 20 books, he had a sideline in journalism, writing for the *Baltimore Sun* and Mencken's *American Mercury*. He makes a cameo appearance in Sinclair Lewis's 1925 novel *Arrowsmith*, about one scientist's struggle to maintain his individuality in the face of large organizations and their money. Inevitably, some colleagues found Pearl insufferable, partly because he dismissed all criticism as the product of minds duller or more conservative than his own. Such a gung-ho approach to life and science led him into blunder as well as triumph. Thinking he had spotted a negative correlation between cancer and tuberculosis in medical data, he injected some terminal cancer patients with TB. The errors in his analysis (although apparently not the experimental treatment) did considerable damage to his credibility and later career. On the plus side, in 1938 he was among the first people to show that smoking is associated with reduced life expectancy.

Pearl believed that an organism's life span was its most important attribute and used his academic freedom to try and work out what determined its duration. In his laboratory he studied the effects of starvation, population density, temperature, and heredity on longevity in fruit flies and on cantaloupe melon seedlings. The plants that grew quickly, he found, died before the slowpokes. Every living thing, Pearl concluded, was born with a certain amount of "inherent vitality," and the speed with which this was consumed determined its life span. "In general the duration of life varies inversely as the rate of energy expenditure during life," he wrote in his 1928 book, *The Rate of Living*.

But these experiments were only a means to Pearl's goal of understanding human longevity. You can't starve humans in the name of science, but in statistics Pearl thought he had the tool that would crack the problem. In 1924 he analyzed the death records of British men of various occupations and concluded that, thanks to the rate-of-life effect, hard labor was the route to an early grave. Coal miners and blacksmiths died younger than lawyers and insurance salesmen, he claimed, even accounting for differences in diet and accidental death. This, he argued, also explained why women lived longer than men— they didn't work as hard. In 1927 he wrote an article for the *Baltimore Sun* called "Why Lazy People Live the Longest."

Pearl never suggested what his inherent vitality might consist of or how it might get used up. We now have a good candidate for a link between metabolism and mortality, but instead of being a good thing that disappears, it is a form of damage that accumulates.

Dangerous Radicals

Oxygen is a mixed blessing, because it is highly reactive. This property makes it perfect for releasing energy from food but also makes it liable to react with and damage the rest of the body's molecules—just as a fire can both warm your house and burn it down. When two oxygen atoms are joined in an O_2 molecule, they are relatively harmless. The trouble starts when the molecule is broken up for use in respiration. Such reactions create free radicals, lone oxygen atoms with spare electrons. It is these spare electrons that make the free radical so reactive and so noxious. The free radical careens around the cell, damaging any protein molecule or DNA it bumps into. DNA damage is particularly bad because it is passed on to the cell's descendents. And when a free radical does find a molecule to react with, it is likely to create more lone oxygen atoms in a chain reaction. This is why dieticians recommend foods rich in small molecules, such as omega-3 fish oils, that mop up free radicals before they can cause harm.

In 1956, Denham Harman, a chemist at the University of California in Berkeley, suggested that free-radical damage could be one cause of aging. About one in every thousand oxygen molecules broken up in mitochondria during respiration produces a free radical. So the faster mitochondria process oxygen, the more free radicals are produced and the more damage is done—providing an explanation for the rate-of-life theory. Studies since Harman's have reinforced the view that free-radical damage is a central process in aging and one of the most important determinants of life span. As a cell ages, it gets worse at doing the things it is supposed to, such as dividing and making useful chemicals, and is more likely to die, attack other cells, or embark on a malignant growth spurt. Old cells shut themselves down or die as a means of preventing such mishaps. Free radicals can replicate many of the effects of aging, mainly by damaging DNA in the cell's genome and the smaller mitochondrial chromosome. Free-radical damage is a vicious spiral—mitochondria produce free radicals and are damaged by them, which makes them more likely to produce free radicals, which increases the damage, and so on. Mice genetically engineered to have more mutations in their mitochondrial DNA show symptoms of premature aging such as declining fertility, curvature of the spine, and hair loss.

Many of the genes linked to life span in the worm *Caenorhabditis elegans*—the animal of choice for many researchers who study aging—affect the animal's ability to deal with free-radical damage. Free radicals are implicated in hazards of old age such as cardiovascular disease, failing eyesight, and neurological disorders, including Alzheimer's disease. The case against free radicals is not closed. Scientists debate how much can be learned about the lives of whole animals from studying cells in a dish, and also about what aging in animals reveals about the process in humans—and we still can't answer Geoffrey West's question of why maximum human life span is about a century. But to quote an article published in *Nature* in 2004, current ideas of how aging works can be summarized as: "It's the free radicals, stupid!"

The Bat Paradox

Metabolism produces free radicals. Free radicals cause aging. The faster you metabolize, the more free radicals you produce and the sooner you die. This does indeed seem like the unity of a great law. But it's not as simple as that. For a start, many things other than metabolic rate influence life span. Contrary to what Neil Armstrong might think, people who exercise do not hasten their deaths—quite the opposite. Exercise can increase free-radical production, but the benefits to the heart, lungs, and vascular system more than outweigh the downside.

Still more troubling for crude rate-of-lifers is that many animals live far longer lives than their metabolic rates alone would predict. The most spectacular examples are bats and birds. A bat smaller than a mouse can live for 30 years or more. Hibernation, during which the animal burns much less energy, might account for some of this, but tropical bats, which do not hibernate, are similarly long-lived. Birds have higher metabolic rates than mammals but much greater life spans. Studies of wild birds have revealed that seabirds can live for half a century, and some parrots are thought to have reached twice that age. Marsupials, on the other hand, have relatively slow metabolisms but short lives. A captive kangaroo will be lucky to see its twentieth birthday. Hyenas, which are of a similar size, have been known to live twice as long. And if you work out how long a billion of your heartbeats would last, you will know that, at least in rich countries, most of us live for much longer than our size would suggest. Even at a healthy 60 beats per minute, 955,787,040 seconds lasts only 30 years. Supposing a strict link between size, metabolism, and longevity, a mammal as big as a human should live only 27 years. This might have been a ripe old age for our prehistoric ancestors, but human hearts and bodies are clearly good for at least several decades more.

The experimental evidence for the rate-of-life theory is also ambiguous. One surefire way to extend an animal's life is to starve it. Rats, mice, worms, and even yeast cells, fed a diet containing all the necessary nutrients but a third fewer calories than they themselves would choose if given free access to food, live about a third longer than those on full rations. Such underfed creatures do have lower metabolic rates—mainly because their body weight has fallen—but many other aspects of their biology change, and it is not clear what contribution slower metabolisms make to their extended life span. In 2004 a team led by John Speakman of the University of Aberdeen revealed that mice with faster metabolisms live *longer* than normal ones. If anyone still held to the simple idea of living fast and dying young as originally proposed by Rubner and Pearl—and by then few, if any, serious aging researchers did—these experiments should have changed their mind. But the links between metabolism, mitochondria, and aging, although more complicated than they once appeared, are still strong.

It now looks as though it's not how quickly you burn energy that influences the aging process but the way that you burn it. Long-lived mice do not simply have faster metabolisms than their shorter-lived counterparts. They also have different mitochondrial chemistry. Mitochondria can do two things with food: turn it into cellular fuel, in the form of ATP, or burn it off as heat. The chemical reactions that turn food into heat produce fewer free radicals than those that produce ATP. Cells have proteins that switch their mitochondria from ATP mode into heat mode, a process called mitochondrial uncoupling. The long-lived, energy-hungry mice in the Aberdeen experiment had a higher level of this uncoupling. Humans use mitochondrial uncoupling to keep warm. Babies have a tissue called brown fat that is stuffed with mitochondria and with proteins that uncouple them, and which disappears around the age of one. And the indigenous people of cold climates have more mitochondrial uncoupling than temperate or tropical groups. This might be part of the reason why cold dwellers have lower rates of neurodegenerative disorders, such as Alzheimer's, but are more prone to diseases of energy metabolism.

Some researchers, including Speakman, are on the trail of drugs that might increase life span by uncoupling mitochondria. We already know some chemicals that do that, but they tend to have unfortunate side effects. One such compound is 2,4-dinitrophenol. Found in explosives and insecticides, this chemical's metabolic properties were discovered during the First World War, when workers exposed to dinitrophenol in munitions factories began losing weight at startling rates—because their metabolic rates had shot up. Dinitrophenol was sold as a slimming drug in the 1930s but was later banned because an overdose makes the body cook itself. Some bodybuilders still use it as a means of crash dieting, and some have died from it. Another chemical that uncouples mitochondria is Ecstasy, or MDMA, which often causes an unpleasant and potentially dangerous rise in body temperature. The drug produces this effect by activating uncoupling proteins in the cells. Together, this evidence raises the thrilling prospect of a pill that makes you thin, high, and ageless, if rather flushed and sweaty.

Bending the Rules

This doesn't explain why some species live so much longer than others of the same size and metabolic rate, but thinking about longevity in evolutionary terms can. Like everything else, how long an animal lives is a trade-off. Maintaining and repairing a body takes energy. There are ways of limiting free-radical damage; for example, cells have enzymes that dispose of free radicals and repair their effects. But making these repairs diverts resources from other processes, such as reproduction. To register on evolution's scoreboard, an organism must survive long enough to reproduce once. After reaching this goal, many species do not bother with a second attempt. Some marsupials, fish, most squid and octopuses, most insects, and all annual plants reproduce only once and then die. By investing everything in one breeding attempt, they can produce more offspring now at the expense of reproducing in the future. Of course, animals from albatrosses to zebras, and plants from grasses to redwoods, invest in keeping themselves healthy enough to breed more than once. They are gambling that this investment will pay off in the long run.

Species are tipped one way or the other by a combination of the ease of reproducing—how much effort they must put into each breeding attempt—and the likelihood of living long enough to breed again. The more effort breeding entails, and the smaller the chances of long-term survival, the more evolution will favor the all-at-once approach. What's a salmon going to do once it has spawned? Swim all the way back out to sea and all the way back in again next year? Forget about it—predators kill most fish before they can complete the journey once. The same reasoning probably explains why male redback spiders will go to great lengths to be eaten by their mates, even somersaulting onto the female's jaws. Females are so hard to find, and a spider's life so chancy, that it pays a male to put everything, literally, into the one attempt and also use its body to feed the female carrying its young.

But not all animals lead such risky lives. If an animal has a good chance of avoiding predators or surviving the next frost to breed again, it might pay for it to slow the aging process and keep the body going. It is striking that both bats and birds have an excellent means of escaping danger—flight. Investing in repairing free-radical damage is a waste if you stand a good chance of ending up in a cat's jaws or a hawk's talons before the week is out, but the chances of this fate befalling a pigeon seem much less than they do a rat. This is probably why a cosseted pigeon can live for 35 years, but a pet rat of about the same size will live only about four years. In a lab dish, mitochondria from pigeons' cells produce less than half as many free radicals as those from rats. Perhaps early humans' social networks and technology cut the rate at which the weather and wild animals picked them off and so led them to evolve cellular systems that slowed the damage of aging. Small rodents, on the other hand, are so likely to die from extrinsic causes that natural selection has never favored a slowing of their aging process.

But no animal can lead a perfectly safe life. And simple probability dictates that the longer you live, the greater the chance of encountering a falling tree, a nasty infection, a poisonous mushroom, a man-eating shark, or a slippery step. No amount of investment in self-maintenance can eliminate these dangers, and this tips the evolutionary balance away from staying healthy and toward reproduction. At some age, depending on life's risks, the value of repair diminishes to the point where the benefits of longer survival become invisible to natural selection. There can also be benefits to breeding early. An animal that breeds at age one year can make copies of its genes twice as quickly as one that waits until it is two. This means that evolution can favor traits that give an animal more offspring early in life but that cause damage later. Flies bred to live longer produce fewer offspring, and vice versa. The same relationship is seen in the genealogical records of the British aristocracy. Bodies age and die because, to evolution, once they have reproduced, they are disposable.

Brian McNab, an ecologist at the University of Florida, has spent decades studying the interactions between an animal's metabolism and its environment. He can probably account for a wild animal's metabolic rate better than any other biologist in the world. Using body size, altitude (which is similar to temperature because higher places are colder), and diet (which is similar to chemical composition), he has explained 99 percent of the variation in metabolic rate for birds of paradise and a group of bats. But he does not think that knowing an animal's metabolic rate reveals everything about it, and, unlike Jim Brown and company, he doubts that simple, universal predictions can be made based on metabolic rate and body size. "Animals have a lot of ways to bend the rules," he says. The relationship between metabolism and aging seems to be the best example of this rule bending. The quarter-power scaling of life span to body size shows that, at the broadest scale, animals live by the ticking clock of metabolism. But the fact that this rule is frequently so poor at predicting the life span of a particular species shows that evolution can favor animals that invest in slowing their metabolic clocks. Nevertheless, without an idea of how energy use and body size influence biology, we would not have realized, for example, that birds live unusually long lives.

Nature's Gamelan

Gamelan is the traditional music of the Indonesian islands of Java and Bali, a mesmerizing sound world built around repeated patterns played on metallophones and gongs. Two nights a week I play in a gamelan orchestra. In a western orchestra each instrument's part is written down and predetermined. Gamelan is not like that. The notation for each piece is minimal, just enough to point a few of the ensemble's instruments in a certain direction. Gamelan notation is more like the rules of a game than a computer program; each musician takes the rules and constructs his or her part based on a set of conventions, handed down from senior musicians, about what one can play in any situation. The rules are flexible. There are often several options for every situation, and every group will interpret the rules differently, so the same piece sounds very different from group to group. Each piece is also flexible: What notes, or even pieces, one plays, how one plays them, and how long a piece lasts are not predetermined but arise from the musical dynamic between group members.

A group of physicists recently concluded that Javanese gamelan was, in terms of rhythm and volume, the world's most complicated music. I was surprised because I have been in workshops where a group of complete novices learned how to perform a piece in only 20 minutes. The complexity of gamelan is an emergent property, of many players following simple rules.

The laws of metabolism are like gamelan music: an underlying structure on which living things can elaborate and improvise. Every place has its own unique conditions and history, and so every species, and population within that species, will be different than the rest, just as every Javanese village has its own musical traditions and conventions. And just as a gamelan group builds a world of music from a handful of notes, nature's diversity takes the form of myriad variations on the theme of turning energy into offspring.

The rules of metabolism become most obvious when you look at many species over a broad size range. The more living things you encounter, the more impressive the similarities and regularities become. This pattern points the way to an understanding of nature that goes beyond the individual—after all, every wild place contains lots of species of lots of different sizes. And once you understand how individuals use energy, and if all use energy in the same way, the patterns seen in forests, flocks, and herds stop seeming like a mass of unconnected examples and start to look like the product of underlying rules.

7   SEEING THE FOREST
FOR THE TREES

Alwyn Gentry worked at the Missouri Botanical Garden in St. Louis throughout the 1970s and 1980s. He specialized in identifying and classifying a group of tropical vines called the Bignoniaciae, or bignones for short. As a graduate student in the late 1960s, Gentry wanted to show that bignones were the most numerous and important tropical vines. For this he needed a way of measuring the forest's plants as a whole. He went to a forest in Panama and laid out a line 50 meters long. Then he went along the line and, whenever he came to a plant that was less than a meter from the line and had a stem that was more than 2.5 centimeters across at chest height, he recorded its size, position, and species. Besides trees his survey included many vines, which can grow thick, woody stems. When he reached the end of the line, he had a long, thin rectangle of forest in which he knew the position, size, and identity of every large plant. He did nine other lines in the same plot, ending with an intimate portrait of 1,000 square meters of forest.

Looking at his data, Gentry realized that he could go beyond bignone advocacy. He had invented a technique for getting a handle on forest biodiversity that required only a few days' work. Ecologists had done plenty of surveying in forests, but no two used the same method, and they tended not to be experts at identifying plants, so there was no global picture of how the species, sizes, and spacing of forest plants varied from place to place. Over the next 25 years, Gentry, a scientist with much energy and enthusiasm, and less patience, put that right. He became an expert at getting hold of small amounts of money to fund short expeditions, and he toured the world collecting what came to be known as Gentry plots. He enlisted his colleagues and students in the project and employed locals and their children to measure, collect, and process samples. When he went somewhere to give a talk, he would make time to survey any nearby forest. In total he gathered data from 227 plots on six continents, from Germany to Chile and Madagascar to Australia, measuring and identifying 83,121 plants. He also carried on his work on plant taxonomy, collecting more than 80,000 specimens for the Missouri Botanical Garden.

Gentry's database showed how forests varied with factors such as latitude, altitude, and rainfall. But the project was always more than an academic exercise. Gentry also worked for the environmental organization Conservation International, and invented his surveying technique partly as a quick and cheap way to get a snapshot of forest diversity, to record the loss of species, and to provide information to plan conservation policy. His surveys helped reveal the impact of people on the natural world. In 1975, for example, he visited the Centinela Ridge in the Ecuadorean cloud forest and discovered more than 100 species new to science. The next year the ridge was cleared for timber.

On August 3, 1993, Gentry was back working in the mountains of Ecuador. He missed his flight out of Quito and chartered a small plane to take him and his team on one more collecting trip. Flying through a cloud bank, the pilot misjudged his altitude, caught a treetop, and crashed. Gentry, the pilot, and one other researcher were killed. Before he died, Gentry broke the plane's windows with the poles he used to collect tree samples, allowing another passenger to escape.

The Way to San Jose

The day after visiting Santa Fe I fly south to San Jose in Costa Rica, swapping the New Mexican desert for tropical green. There I am to meet Brian Enquist and his team, to join them as they do fieldwork and learn about forest ecology. I think that driving my rental car from the airport to the hotel where we are to meet will be a cinch: San Jose has a grid system of streets. What I don't know is that hardly any of them bear signs. I inch my way across town a block at a time, driving until I know where I am, then reorienting and setting off on a new bearing. I feel rather self-conscious in my white, soccer mom's SUV, which might as well have a flashing neon sign that says TOURIST on the roof. When I despair, I put things in perspective by reminding myself that there are journalists driving around Baghdad. After driving once in each direction down every road in town, not excluding one-way streets, I reach the hotel.

We rendezvous and leave town for the forest. After one stop to buy beer and two stops to look at plants, we arrive at our destination, an ecotourist lodge on the banks of the Rio Savegre, catering to bird watchers from North America and Europe. Several species of humming-birds zoom in and out from the hotel's nectar dispenser. Their names paint their pictures: Scintillant, Purple-Throated Mountain Gem, Green-Crowned Brilliant. The day after our arrival we drive up into the cloud forest. As soon as we get out of the cars, we hear the whooping call of the bird everyone comes to see. Sitting on a branch above the track is a male Resplendent Quetzal. It looks like a bird invented by the Costa Rican tourist board: the size of a large pigeon, scarlet on its breast and green everywhere else. It is hard to imagine that more vivid shades of red or green exist anywhere else in the world. The quetzal flies away, it's 2-foot-long tail undulating behind it, and we turn our attention to the trees.

This is my first time in a tropical forest. The first thing I notice are the big trees—40-meter monsters that burst through the canopy, with trunks more than a meter across, propped up by buttress roots that double this breadth. In this forest all such trees belong to various species of oak. Then I notice ways of being a tree you never see in the British woods I know—tree ferns, palms, and bamboo. After that I notice how each tree supports a city of other plants. Every branch and trunk is swathed in moss and studded with orchids and bromeliads. These plants-on-plants are called epiphytes. Even the leaves have a film of other plants growing on them. Lianas, such as Gentry's beloved bignones, snake up around trunks. Other dangling stems look like they belong to vines but in fact come from species that germinate on high branches, like epiphytes, and send their roots downward to make contact with the ground. Over decades or centuries, these can encase and kill their host tree: The strangler fig is the best-known example of such a strategy.

But to really appreciate the diversity of this forest, you need to be an expert botanist. Luckily, our party includes Brad Boyle, a grad student of Gentry's when he was killed and now a postdoc in the Enquist lab. Brad knows a remarkable number of ways to identify a tree. He looks at the curve and shape of a leaf, whether it is hairy or smooth, and how the leaves are arranged on the stem. He looks at leaf stalks. He pulls out a small magnifying glass and looks for translucent spots in the leaf, which botanists call punctations, and the remnants of structures that protect the young leaf, called stipules. Sometimes, he is the only one who can interpret the presence, or absence, of these features. He crushes leaves and learns from their smell and texture. He examines the color and texture of bark, and slashes trunks with a blade to see the color and smell of the wood beneath and any sap that oozes out. This technique teaches me how to spot trees in the nutmeg family, because their wood smells like the spice; it also teaches me that the wood of the genus *Croton* smells like gari, the preserved ginger served in sushi restaurants. Sometimes Brad pops a leaf in his mouth and chews for a few moments before announcing the plant's identity. *Drimys* leaves, I learn, taste bitter and make your mouth go numb. Within a few hundred meters of our parking spot, Brad tracks down more than 40 tree species.

Over the next three days we complete our own Gentry plot—the group calls it a gentraso—in the cloud forest alongside the Rio Savegre. We lay out 50-meter lines and move along them, measuring and recording. Not even Brad can identify every tree right then and there, so each species gets a temporary name, such as "saw leaf," "donkey ears," or "son of donkey ears," and we collect a sample for later identification. Nate Svensson, a graduate student in Enquist's lab, bores small

Brad Boyle in the Costa Rican forest.
Credit: Jason Pither.

cylinders of wood out of tree trunks and slips each sample into a drinking straw for storage. Back in the lab, he will measure the density of the
wood. I had expected the forest canopy to cover the sky, and it does,
but I am surprised to find that every space between the treetops and
the ground is filled with plant life. Moving through the forest is like
swimming along the bed of a sea of vegetation—swimming uphill, that
is, as the plot lies on a 45-degree slope.

Each day's work yields half a dozen black plastic sacks filled with
plant samples. In the evening, our room at the lodge becomes a temporary lab. Brad sits at the head of the table and pays out specimens for
the rest of us to slip between sheets of newspaper. These will later be
pressed properly, filed in a museum collection of plant specimens,
called a herbarium, and eventually identified. On each sample is
written Brad's name and a number that marks the running count of
specimens he has collected in his career. The first one made on this trip
is 7,500; the last is 7,736. (Among plant collectors this count is a
measure of status. The very few who get to 100,000 obtain a higher
plane among their colleagues, a botanical nirvana.) Each sample taken
from a plant is split into four smaller bits for pressing. Two copies will
stay at institutions in Costa Rica and one will go back to Arizona. The
fourth is for tricky cases that may need to be sent to a world expert in a
particular plant group, perhaps at the Missouri Botanical Garden, the
Kew Gardens in London, or New York's Botanical Garden. It's satisfying to think that the specimens I am making will be good for centuries,
a permanent record of what was growing at a particular place and time,
a brick added to the botanical edifice. Perhaps—who knows?—they
will end up in a herbarium rubbing shoulders with plants collected by
Darwin or Linnaeus.

Meanwhile, Drew Kerkhoff, another postdoc in the Enquist lab, is
analyzing the thickness and chemical composition of leaves. He spends
the evening with a hole puncher, perforating leaves to create uniform
disks of tissue for later analysis. To record the area of unpunchable
leaves, such as fern fronds, he uses a modified image scanner, a device
that reminds me of the modified paint roller Samuel Brody used to
measure the surface of cows. Jason Pither, another postdoc, completes
the team. He is not a forest ecologist; at the moment he is working on

diatoms, microscopic water plants. But he can use a hole puncher as well as the next person, particularly when the next person is me.

The team is a blend of complementary expertises: Brad is a naturalist, Drew primarily a theoretician, and Brian somewhere between the two. Their personalities are similarly interlocking: Brad gives out a Zen-master air, and from his anecdotes seems to be omni-competent. (One that particularly sticks in my memory is best summed up as "The time the engine fell out of my VW camper van, but I fixed it with a chain and a plank of wood and drove on to Guatemala City.") Drew is sardonic and makes most of the jokes, and Jason absorbs most of them. Jason also keeps things running smoothly. He is usually the first to plunge into the forest trailing a tape measure, and the first to pack up equipment and gather up rubbish, and to check that everyone else is OK. Brian is in charge, but I don't remember seeing him, or anyone else, give a direct order on the whole trip. The dynamic is a bit like a jazz band. Everyone is simultaneously working for the group and doing his own thing, without an obvious leader. Sometimes we drink beer and gossip about science, scientists—whose ideas make sense, whose don't, who's collegiate, who's paranoid—academic life, and which rock groups should be sent to the bottom of the sea in a sub-marine. Other nights we work heads down, in near-silence. It's not exciting, but it requires just enough concentration to stop your mind from wandering. And there are plenty of opportunities to make dumb mistakes. My heart sinks when I realize I have been labeling specimens wrongly, but fortunately the error is correctable, and no academic careers are ruined. Over the course of an evening, the plastic bags of leaves and branches are transformed into piles of specimens flattened and wrapped in newspaper, like bundles of freshly ironed shirts.

After finishing the plot at Savegre, we go north to do another gentraso, in a forest just shy of the Nicaraguan border. We stop once to buy beer, once to drop specimens off at the Costa Rican National Biodiversity Institute in San Jose, where they will be dried and mounted, and once at a stationery store to buy an extra hole puncher, so Drew can get more help—the punchers are having trouble keeping up with the pressers. As I drive, I speculate about what sort of power law best describes the size of the potholes on the Pan-American

Highway. Our destination is the Volcan Cacao biological station, half an hour in a four-wheel-drive vehicle from the nearest road and half an hour on foot from that track. We arrive after dark and walk through a forest loud with cicadas and bright with fireflies. Unlike Savegre, Volcan Cacao doesn't have a bar, buffet meals, or hot showers. It does have cold showers, in which a colony of several hundred wasps has set up home, and wooden bunkhouses with corrugated iron roofs. At night the noise of the rain makes me dream of applause. It also has a veranda from where you can watch the sun go down over the hills and into the Pacific. The morning after our arrival we hike from the station up into the forest, crossing paths with columns of marauding army ants. The trade winds blow continuously from east to west over Cacao, washing a river of cloud over the trees around the mountain's summit. The clouds deposit a small pool of water into the center of each bromeliad and turn every clump of moss into a sponge; the forest is like an aerial bog.

Just about anyone could do a gentraso. Our most sophisticated pieces of equipment are long aluminum poles with clippers at the end, of the sort Gentry used, to snip leaves from high branches, and a forester's tape measure that, when wrapped around a tree trunk's circumference, gives a reading of its diameter. It's not exactly the Hubble Space Telescope. Nor will our measurements have a microscopic precision. Vines and moss growing around the trunk can inflate the measure of its width. Chest height is set as 1.3 meters up the body but is then judged by eye, so the measurement varies between trees and between people. Measurements of tree trunk breadth are converted into a reading for the volume of the whole tree, because a tree's height and width are correlated, but such calculations are approximate. It's impossible to lay out the 50-meter lines perfectly parallel to one another or perfectly straight. At one point, while running out a line at Cacao, and already wet and muddy from pushing through the forest, I fall several feet into a gully. When I climb out, I change course to avoid this happening again. It's said that Gentry would change course to take in any interesting tree that caught his eye. There is some subjectivity to where the 1-meter boundary falls for a tree being in or out of the plot. This expedition seems more like old-fashioned natural history than a route to a grand theory of biology.

And at first glance it looks a forlorn hope that anyone could understand the forest. It's a green riot, a place of disorder and confusion, a jungle. The only possible statement seems to be that if a thing can grow, it will. But as the measurements build up, pattern begins to emerge. The number of stems we measure on each line is fairly constant. At Savegre each 50-meter run brings in about 30 plants—the range is from 25 to 37. Cacao shows a similar consistency, but the lines there have upward of 40 individuals in them. The most crowded have nearly 60. As Brad says, "It's a busy forest." But whereas the biggest trees at Savegre have trunks more than a meter wide, there are no such giants at Cacao. The biggest stem we encounter is less than 60 centimeters across. So the Savegre cloud forest has fewer, bigger trees, whereas the forest at Cacao has a greater number of smaller ones.

It would, of course, be rash to make any firm conclusions about the world's forests from visiting two places in one Central American country. But while the techniques of measurement and collection have changed little in the past couple of centuries, computer power and methods of data analysis have. To do a statistical regression of 100 data points by hand, finding the equation that best describes the pattern in the data, requires a day or two of laborious calculations. A computer can do it before your finger comes off the return key. The Internet makes it much easier to find and share data from different sources. Through a combination of their own fieldwork and ferreting in the scientific literature, the Enquist lab has collected a set of gentrasos three times larger than that collected by Gentry himself. And the more measurements you collect the more strongly order emerges.

How to Build a Forest

Think about the differences between Cacao and Savegre in the population densities and sizes of trees. It's obvious that any area will support fewer big individuals than small ones, because big plants (and animals) need more space and more resources. This scenario is played out in time as well as between different places. In a forest after a fire or in a gap created by a fallen tree, many saplings will sprout—many more, in fact, than the number of full-grown trees that previously lived on the spot. For a while there will be enough resources for them all to coexist. But as they get bigger, they will start to crowd and compete with one another, and most will die. Gardeners and foresters plant many more seeds than their plots can support and then thin out the seedlings as the plants grow, leaving room for some to become large. Something similar happens in natural communities. The same process is at work in any place where there is competition for space and resources, in any place where life is a struggle. In other words, everywhere.

Botanists, in a nod to horticulture, have called the way density changes with size self-thinning. Not surprisingly, the relationship between size and population density is best described by a power law. The questions are: What form does this power law take? As individuals grow, by how much must their density go down? In 1963 a group of Japanese ecologists, led by Kyoji Yoda, concluded that the slope relating the logarithm of the size of plants in a plot to the logarithm of their density was $-3/2$. Other studies, on species ranging in size from moss to trees seemed to confirm that both within and between species this power law reflected a precise relationship between plant size and population density. Self-thinning became a prominent part of ecology's intellectual landscape and the exception to the rule that botanists don't do theories. The pattern was seized on as a rare example of a precise mathematical expression of nature and described as "the only generalization worthy of the name of a law in plant ecology."

Like Kleiber's law, it wasn't clear how the mathematical pattern could be explained. The most popular idea was based on a geometrical argument similar to Rubner's surface law. The area of a plant's canopy—the amount of ground it occupies, which will control the number of plants you can pack into an area—is proportional to the square of that canopy's radius, in the same way that the area of a circle equals pi times its radius squared. The whole plant's volume is proportional to that radius cubed, in the same way that body mass is proportional to the cube of body length. A quick algebraic rearrangement therefore shows that plant volume is proportional to the 3/2 power of canopy area. The inverse of this number—the $-3/2$ power of each plant's volume—would describe the number of plants of a given size that an area can support and how that number would decrease as they got bigger. But, like the surface law, self-thinning did not stand the tests of academic scrutiny. In the mid-1980s it was shown that Yoda's team and those who followed had made mistakes in the way they compared plant size with population density. These mistakes made the relationship between the two properties look stronger than it really was. When the data were reanalyzed, no clear-cut pattern relating size to population density could be seen. By the mid-1990s most ecologists had given up on self-thinning.

In their metabolic model, West, Brown, and Enquist saw a new way to tackle the question. Instead of dividing up space, population density could be a consequence of the way that plants divide up energy. If there is a fundamental unity in the way individuals use energy—if everyone is playing the same game by the same rules—the path from the diversity of individuals and species to the patterns seen in living groups becomes clear. The fractal theory explains what happens to energy as it pours into an individual organism. It seems to apply to all organisms. So an ecosystem becomes a network of networks: Energy pours into it, and flows through all its component organisms depending on how large, competitive, or lucky they are. From this angle, ecology becomes a kind of meta-metabolism.

The network theory of metabolic rate predicts how tubes should be packed into a body to spread energy through it as efficiently as possible. Its ecological extension predicts the best way to pack trees into space. Sunlight is so valuable—and trees are so good at getting it—that by assuming trees use energy optimally, we can work out how a forest ought to spread out. Body size controls this spacing pattern. Just as elephants need more food than mice, an area will support fewer large plants than small ones. But just as an elephant's cells burn energy more slowly than a mouse's, so we must take the relatively slower metabolic rates of large plants into account when working out how many of them can fit into an area.

Metabolic theory predicts that the population density of trees should decline with their increasing size so that the total metabolic activity of the plants remains constant. In 1998, West, Brown, and Enquist published a paper showing that, for plants from herbs to trees a billion times larger, the number of individuals in an area declines proportional to the –3/4 power of their body masses. That is, population density falls at the same rate as metabolic rate rises. A 100-fold increase in plant mass leads to a 30-fold decrease in population density; a square meter of ground can support 200 plants each weighing 10 grams, or seven plants if each weighs a kilogram. Data on plant population density support this prediction: At the extremes, a square meter supports about a million duckweed plants, each weighing one ten-thousandth of a gram, totaling 100 grams of plant matter, or 1 percent of a sequoia tree weighing 10,000 kilograms, totaling 100,000 grams.

As well as predicting how many trees of a certain size can fill an area, metabolic theory predicts the range of tree sizes in a forest and how those trees space out. Large trees are few and far between. In Savegre and Cacao no survey line in any gentraso had more than one truly hefty tree, with a trunk more than 40 centimeters across. But each line contained many trees that just make it over the 2.5-centimeter cut off point. Look at lots of trees and a scaling law emerges: The wider the trees, the greater the space between them. So if you are standing by a tree you can predict how far you will have to walk before you find a tree of the same size. The bigger the tree, the longer the journey. The forest consists of a few giants, keeping their distance from one another, and many small, more gregarious trees filling the gaps between them. To see why this is so, imagine you had to fill a crate with balls of different sizes. There would be room for maybe only two or three basketballs. But the gaps between them would leave room for perhaps five or six volleyballs. Between these volleyballs you might be able to insert a dozen baseballs and still have room to cram 20 golf balls into any remaining gaps. The amount of energy in the forest—representing the size of the crate in this analogy—can accommodate only a few really big trees. But many more small trees can squeeze into the gaps between them.

The trees of different sizes are like a network for using the sun's energy, and trees fill up space in the same way as transport networks. The same mathematics that predicts how to design a blood system also predicts the numbers of each size of tree, with the number of large, medium-sized, and small trees following the same pattern as an aorta that leads to several arteries and many capillaries. It's another example of how, by assuming that life is structured so as to grab as much energy as possible, you can predict what large-scale order will emerge.

Animals in Space

In their efforts to relate body size to population density and metabolism, botanists were once again following in zoologists' footsteps. In 1981, John Damuth looked at the data on body size and population density for 307 species of herbivorous mammals. He found that population density declines as the –0.75 power of body size. Because metabolic rate increases as the 0.75 power of body size, this means that all animal populations extract approximately the same amount of energy from an area, regardless of how big they are. This is just what West, Brown, and Enquist found for plants nearly 20 years later—that 1 hectare of grassland uses the same amount of energy as 1 hectare of forest. The principle has become known as the energy equivalence rule. In a later study of more than 600 species, Damuth found that the same relationship held for animals ranging from a mite, *Chamobates schützi*, which weighs 0.0000083 grams, to the African elephant, which weighs 2,900 kilograms. A square kilometer's worth of mites—about 4 billion animals, weighing about 33 kilograms—uses approximately the same amount of energy as the lone elephant that the same area could support. Similarly, the population density of parasites in animals declines with the –3/4 power of the host's body size. One of the advantages of being large is that, pound for pound, bigger animals have fewer parasites than small ones because the parasites, just like their host's cells, must make do with a slower supply of energy.

But although energy equivalence worked well for herbivores, plants, and parasites, it did not seem to apply to animals higher up the food chain. For them population density declined more steeply with size than metabolic rates would predict. Instead of energy use remaining constant, the brute mass of carnivores in an area seemed not to vary—a square kilometer could house the same mass of lion flesh as weasel flesh. This was puzzling because the smaller weasels, with faster metabolic rates, ought to require more energy, but the data seemed robust, and by the late 1990s it looked as if energy equivalence had no claim to be a general biological principle.

But in 2002, two zoologists, Chris Carbone and John Gittleman, found a way to bring carnivores back into the metabolic fold. Their trick was to stop thinking about animal populations in terms of the land they needed, and instead to think about the amount of food they needed. For plants and herbivores, land and food are much the same thing. As much light falls on a hectare of grassland as on a hectare of forest; grazing rabbits and grazing wildebeest encounter grass in much the same way. But carnivores are more idiosyncratic. They live in lots of different places and hunt in lots of different ways, and they tend to specialize in hunting certain prey. In particular, small carnivores tend to focus on eating invertebrates—like a hedgehog snuffling for worms—which are much smaller than them, whereas large ones tackle vertebrates that are often nearly their own size, such as a cheetah in pursuit of a gazelle.

The switch between the two styles of hunting occurs around a dozen kilograms. For example, the European badger and the American coyote both weigh about 13 kilograms. But the badger feeds mainly on invertebrates such as worms and beetles, whereas the coyote prefers mice and rabbits. Anything larger than a coyote will likewise hunt relatively large prey. Worms are spread evenly, like plants, but mice and rabbits are spread much more thinly, so coyotes must spread out more thinly than badgers. This means that the population density of carnivores takes a sudden drop around the 12-kilogram mark, and so trails off more steeply with increasing body size than does the density of herbivores, which do not show the same switch.

When Carbone and Gittleman looked at carnivore populations in terms of the amount of *prey* available, rather than the amount of land, energy equivalence was restored. For species ranging from the least weasel, which weighs 140 grams and needs less than a square kilometer of land, to the polar bear, which weighs 310 kilograms and needs 50 square kilometers or more, the population density of carnivores falls at a rate proportional to the $-3/4$ power of prey density. This energy equivalence means that every kilogram of meat eater needs about 111 kilograms of meat to keep it going (the ratio remains constant despite changing body size because, although a kilogram of weasel burns fuel more quickly than a kilogram of bear, and so you might expect it to need more meat, the species that weasels hunt are themselves small and so grow and reproduce more quickly than bears' prey). Carbone and Gittleman's finding also shows that discovering general patterns requires local knowledge. Without an understanding of the feeding behavior of different predators and the population densities of their prey, the rules describing carnivore numbers would have remained obscure.

The match between metabolic rate and population density also explains why any area can support a greater mass of reptiles than mammals, which in turn live in more dense populations than birds. As an animal's body temperature, and so its metabolic rate, goes up, its population density must go down. This is why, as Carbone says, "we have fisheries but not birderies"—and there's no point trawling for leopards. Many economically important fish, such as cod and salmon, are predators, but their relatively slow metabolisms allow them to live in numbers that would be impossible for a group of warm-blooded predators to sustain. The population density of predatory theropod dinosaurs, the group from which birds evolved, seems to have been much lower, relative to the density of their prey, than other predatory dinosaurs, hinting that the theropods might also have been warm-blooded.

As we have seen, conservationists use the links between body size and the pace of life to work out which species are most at risk of extinction. Likewise, they can use the links between energy and land to work out whether, and how, species could be saved. A nature reserve cannot protect an endangered rhino species unless it contains enough food to support a breeding population of rhinoceros (working out how many rhinos make a self-sustaining breeding population is a whole other issue). The patterns in population density give another rule of thumb. You can get a rough idea of how many animals of a given size could live in the area you are proposing to make into a reserve. This would be simple for herbivores and a bit more complicated for carnivores, as you would also have to measure the population density of the predator's prey species. The rules also give a guide to how well conservation measures are going. Carbone is working to conserve the Sumatran tiger, which hunts mainly wild pigs; an area of forest with much less than 1 kilogram of tiger for every 111 kilograms of pig flesh might be a sign that something else, such as the effects of inbreeding, is keeping tiger numbers down. Macroecological patterns are only a starting point: There is still a lot of variation in population density around the energy equivalence line. But even a crude prediction is better than having no idea of what to expect, and, while tiger conservationists might have the time and resources to do detailed studies, anyone aiming to monitor, say, all the carnivores in a large area of forest or savannah is going to have to take shortcuts.

There is another aspect to how animals use space beyond their population density. Unlike plants, animals move about. Each animal roams over an area of land called its home range. Not surprisingly, big animals have bigger home ranges: One elephant covers nearly 30 square kilometers of land; the same area could accommodate 10,000 mice. But home-range size is not proportional to body mass raised to the power of 3/4; instead, range size rises at the same rate as body size. For some reason, large animals use more land than would be expected based on their energy requirements.

Many animals secure energy by defending territories, as either individuals or a group. They patrol their ranges and repel intruders. The reason that large animals need proportionately larger territories, metabolically speaking, is because they have a harder time defending them. The distance traveled by an animal in a day is proportional to its body size raised to the power of 1/4. The amount of food it needs is proportional to the 3/4 power of its body mass. So the amount of food needed increases with body size more quickly than the area covered. A mouse travels 110 meters in a day, more than adequate to cross its entire territory. An elephant manages 2 kilometers. Large animals therefore find it harder to defend their borders, because they are less likely to bump into an intruder. They must put up with more incursions and share more of their territories and resources with their neighbors. These intruders eat, so each territory holder is forced to roam over a still larger area to find enough food—which is why home range size increases with body size more quickly than metabolic rate.

To express this relationship mathematically, Carbone, working with a team of other ecologists that included Jim Brown, borrowed the equations that physicists use to predict the frequency of collisions between the randomly wandering particles in a gas. Applying this model to animals, incorporating factors such as population density and mobility, they predicted that the relationship between an animal's food requirements and the distance it must travel to get that food would be linearly related to body mass: a good agreement with real-world patterns. The problem that large animals have in defending their territory is probably one explanation of why large animals, such as zebra and bison, tend to live in groups but why, to quote William Calder again, one never sees a herd of mice. When defending a territory becomes unfeasible for one animal alone, it is a better to tolerate company than rush around trying to fight it off.

This is a static view of populations: a snapshot of numbers and density, frozen in time. But animals and plants aren't like this, of course. In 1798, Thomas Malthus noticed that animals seemed capable of increasing their numbers at an exponential rate. A female rabbit produces a litter of six pups. If all six manage to avoid foxes and find enough food, this female's three daughters will each produce six young, creating 18 more rabbits. These will then give birth to 54 rabbits, then 162, and so on. Each generation will contain three times more rabbits than the last. And yet in the long term, most populations, of rabbits and everything else, must maintain almost constant numbers; otherwise they would either go extinct or cover the earth. Malthus argued that the world is not covered in rabbits because famine, disease, winter, and such keep their numbers in check. Famously, he also argued that the same would happen to a growing human population. Charles Darwin realized that this overproduction provides the raw material for natural selection. Since then, life's capacity for Malthusian geometric population growth has been recognized as one of nature's few universals.

We saw in the previous chapter how metabolic rate controls the growth of individuals. For some species, such as bacteria, the link between individual and population growth is obvious: They are the same thing, because new cells are also new individuals. For others it is more complex: Mothers might invest resources in eggs, fetuses, and parental care. Males might help care for their young, but most do not; instead they put all their reproductive efforts into finding mates. Nevertheless, these are all decisions about what to do with energy, and the rate at which an organism can turn resources into offspring depends on the rate at which it uses resources itself—in other words, on its metabolic rate.

It ought to come as no surprise that the maximum population growth rate—the number that defines each species' Malthusian potential—depends on body size, and that populations of large animals grow more slowly. Like cellular metabolic rate, the maximum rate of population growth for each species is proportional to the −1/4 power of its body size. Temperature is also important. Fish populations in polar seas increase more slowly than those in tropical waters. Together, size and temperature can account for about 95 percent of the variation in population growth rates for organisms from plankton and algae up to large vertebrates. As well as controlling how numerous a population of plants and animals can get, metabolic rate controls how quickly it will get there.

The One Forest

I am no Brad Boyle, but if I was transported at random to some wooded part of the earth, even I could guess whether I was more likely to be in, say, Colombia or British Columbia. The types of trees in a place, and the number of species, vary hugely. The plots in Gentry's data set contained between 2 and 275 species. Why this is so is a puzzle, of which more will be said later. But by analyzing Gentry's original data set, Enquist and Karl Niklas were able to show that, in their large-scale structure, all forests are essentially the same. So maybe the gentrasos really are ecology's equivalent of the Hubble telescope, because they give a view of nature that is of unprecedented scope and depth.

All forests, for example, follow the −3/4 relationship of individual size to body size. This is true of the trees in a single Gentry plot, an entire continent, or the whole world. And all forests contain approximately the same number of individuals and quantity of plant tissue, or biomass. There are the same number of trees (about 400) and the same amount of wood (about 35,000 kilograms) in 0.1 hectares of forest in Amazonia or Alaska. Regardless of the species that live there, natural selection, limited by the constraints of energy and allometry, has pushed all forests up against the same physical limit of the size and number of individuals that a place can support.

The Gentry plots also revealed that no type of forest has a monopoly on tall trees. The Amazonian rain forest, Australian eucalyptus forests, and North American conifer forests all contain trees about 100 meters tall. But why stop there? From a ridge on Volcan Cacao I can look out across the treetops, undulating but level—the surface of the sea in which I have just been thrashing about. This evenness is commonplace but also strange. Each tree, after all, has a powerful incentive to get taller, as it could then grab a greater share of sunlight. Why do forests have a canopy? Why aren't there some trees as tall as the Eiffel Tower?

Part of the answer will, like life span, be a question of life's risks, with storms, or rot from within, bringing down any tree before it got fantastically large. Partly the answer will be biomechanical, to do with the problems of supporting large bodies. But none of these factors offer an obvious explanation of why, in all the world's forests, there is no tree 200 meters tall. But the fractal theory can explain this by showing how the effect of size on transport networks puts an upper limit on tree height. This upper limit is set by the fact that trees rely on water evaporating from their leaves to draw water into their roots, and also by the tapering of the xylem vessels that carry this water.

The width of the narrowest xylem tubes—the leaf stalks that do not vary between species—is fixed. But like animal blood vessels, the tubes' maximum width can vary. Tall trees need wider xylem tubes just as large animals need wider aortas. But evaporation is not as good as pumping at driving water down a pipe, and if xylem vessels got too wide the evaporation from a tree's leaves would be too weak to draw water up from its roots. So the vessels' maximum width is also fixed.

Perhaps if xylem vessels tapered infinitesimally slowly between these two limits trees could become massively tall. But this is not an option. The rate at which xylem vessels taper is set by the need to provide every leaf with an equal flow of water. The physics of fluid flow means that if xylem vessels tapered too slowly a tree's lower leaves would get all the water and its upper branches would perish. So the maximum height of trees corresponds to the length of the xylem tube that bridges the thinnest tubes at the treetop and the fattest possible tubes at the root, while tapering at the right rate. This bridge can reach about 100 meters in length—corresponding to the tallest real trees— before it collapses. Looking down on the treetops and thinking of this concept, I imagine the canopy as a solid, immovable layer of invariant terminal units and the forest growing downward, like a carpet of strangler figs, with shoots and trunk becoming ever broader until they reach the limits of their capabilities and hit the ground.

The fractal theory can predict a maximum tree height, but it can't predict why trees reach the height they do. The Cacao forest canopy, and that of most forests, isn't 100 meters tall. What sets forest canopy heights is still a mystery. One danger of wide xylem vessels is that they are prone to develop air bubbles, cutting off the water supply to the leaves. This may cramp trees' ability to get tall. The wetness of the environment is probably also a factor. Trees need the air around their leaves to be drier than the soil around their roots, otherwise water will not flow through them, and they will not be able to photosynthesize and grow. In very humid places, such as cloud forests, this flow will be slow, which might prevent trees from reaching their full potential.

Meta-Metabolism

The gentrasos are not the only large data set to reveal uniformities among the world's plants. You can't stick a plastic hood over a forest to measure the flow of oxygen and carbon dioxide through its trees, but you can catch its breath in other ways. In the mid-1990s, as global warming rose to the top of environmental scientists' concerns, researchers began to wonder how to study the effects of climate change on ecosystems. Would warmer conditions and more carbon dioxide fertilize plant growth, for example, or would soils become drier and stunt growth? Researchers' solution was to establish monitoring towers that would give an hour-by-hour picture of the conditions in an ecosystem, such as the temperature and the levels of carbon dioxide and water vapor. FLUXNET, as it is called, now numbers 140 sites in five continents, covering all sorts of forests, wetlands, grasslands, agricultural ecosystems, and tundra. Such projects mark the birth of Big Ecology, large international teams of researchers collecting stupendous amounts of information on a global scale (although the cost is still minuscule compared to space missions). And besides being a canary for the effects of climate change, FLUXNET was just what ecologists such as Enquist needed to measure ecosystem metabolism.

The results from FLUXNET confirm some basic truths about metabolism, showing, for example, that forests respire more quickly in warmer weather. FLUXNET data also match the metabolic theory's predictions of energy equivalence. The amount of respiration does not depend on the amount of living plant matter in the ecosystem. It is the same for temperate and tropical forests, grasslands, and soybean plantations—corrected for temperature, every plant-covered hectare of land takes in the same amount of oxygen and produces the same amount of carbon dioxide. This makes sense considering how size affects metabolism. Adding biomass seems like it should increase the amount of respiration going on. But if you do this, you end up with fewer bigger individuals, which must respire at a relatively slower rate. Although the metabolism of individuals depends on their mass, the metabolism of ecosystems doesn't. In every place, individuals, regardless of their size, have—as energy equivalence predicts—evolved to grab hold of as much energy as they can and use it as efficiently as possible.

But forests also do some things not predicted by the metabolic models. The pattern of respiration between forests does not reflect the average temperature—they all seem approximately similar. For any given temperature, temperate forests can process energy between three and six times more quickly. This is why Drew Kerkhoff has been punching holes in leaves.

Moving from the equator toward the poles, and so into colder environments, trees' leaves become steadily richer in phosphorus. Leaves' nitrogen content, on the other hand, remains constant, so the nitrogen-to-phosphorus ratio becomes biased toward phosphorus.

This change seems to allow temperate trees to compensate for their cold environment with relatively accelerated growth. The amount of new wood that each hectare of forest makes each year does not vary from place to place. Phosphorus is a critical component of the cell's protein-making machinery; temperate trees might grow faster thanks to increased investment in such machinery. Presumably there is also a cost for this increased growth that prevents tropical trees, which do not need to cram all their growth into a short summer, from investing in leaves rich in phosphorus.

The One Tree

After leaving Volcan Cacao, we drive toward the coast and nearby Santa Rosa National Park. The park was established in 1971, before which most of its land was used for cattle ranching. By stopping grassland fires and allowing horses and cows to spread the seeds of trees that once relied on large, but now extinct, mammals such as ground sloths, to do the job, Costa Rican conservationists and their U.S. colleagues have helped the trees grow back. The result is a very different type of ecosystem than that of a cloud forest. We are only 20 kilometers from the Cacao plot, but none of the same tree species will be found here. It's the dry season, the temperature is above 30°C, and most of the trees have shed most of their leaves in an effort to conserve water. There are few vines, and the space beneath the canopy is much more open, making it easier to move around the forest. If you squinted and tipped a bucket of cold water over your head, you could almost believe you were in an autumnal English woodland. Except that, instead of moisture-loving mosses, the most obvious branch dwellers are cacti. And the acacia bushes have fierce ants living inside their hollow thorns. And ctenosaur lizards several feet long scuttle out of the road as we approach.

Enquist has been coming to Costa Rica since 1992; in total he has spent more than two years in the country. Most of this time was passed at Santa Rosa, and most of the time at Santa Rosa was spent working on a plot in a part of the forest called San Emilio. The ranchers moved out of here more than a century ago, making this one of the oldest forests hereabouts, with many mature trees. In 1976 another U.S. ecologist, Steve Hubbell, set up a permanent plot for studying the forest here. A gentraso is a snapshot for comparing forests in different places—you can't go back to the same forest and be sure you're measuring the same trees. In San Emilio you can. A grid of stakes divide the forest, allowing every tree's position to be mapped. The larger trunks have metal tags with identification numbers. Unlike a Gentry plot, this system allows the forest to be followed through time. As a graduate student, Enquist inherited the San Emilio plot and spent most of 1994 resurveying the 17-hectare site. With the help of his wife and the occasional visiting undergraduate, he recorded the position, size, and species of every tree with a trunk more than 3 centimeters wide. He measured about 3,000 plants of 45 species. Like a scientific Davy Crockett, he knows every tree in this forest and, giving us a tour of his favorites, is obviously excited to be back. "I'm getting misty-eyed," he says, only half-jokingly.

Gentry plots reveal that a common structure unites forests in different places. The San Emilio surveys showed that the same is true in time. In the interval between the two surveys, the climate became drier, causing a change in the species living in the plot. But there was no change in the size or structure of the trees: The size of the biggest trees remained unchanged, as did the number of trees of each size, and the population density. And the progress of the 2,283 trees that did survive the two decades between the first and second surveys revealed that a unity underpins the biology of different tree species as well as of forests as a whole.

Plants have different life histories, just like animals. Some tree species shoot up quickly; others are slow growing. But in San Emilio, although the sizes of individuals changed at very different rates, their mass did not. All of the trees added biomass at the same rate. In the now-familiar pattern, growth rate was proportional to body mass raised to the power of 3/4, so small trees grew relatively more quickly than big ones. A tree's growth rate depends on its size, not its species, and there are really no slow- or fast-growing trees.

What does differ between tree species is the density of their wood. As well as measuring tree diameter, Enquist took his own set of wood cores; Nate Svensson is extending this data set and might do a third survey of the San Emilio plot. Some trees have wood like poly-styrene; others are so dense they sink in water. Enquist broke several drill bits trying to bore into the toughest trunks. It's another case of trade-offs, this time of wood quantity against quality. If a tree wants to gain height quickly, it can make less dense wood. This seems like a good idea—it can overshadow its neighbors, and win the race to grab a spot in the canopy. Such fast-expanding trees tend to be the first to fill the gaps left by fallen trees. But there is a price to be paid. Speedy growers sacrifice sturdiness in their race for height and have shorter lives than trees that make dense wood. The slow-and-steady growers, grinding out their existence in the shadier spots, are in it for the long haul—tougher, better able to withstand storms and disease, and longer lived.

What applies to wood also applies to leaves. Some evenings I wield the hole puncher. Handling leaves, I learn that some species have foliage like tissue paper; others are more like cards. The same costs and benefits apply: Flimsy leaves are quick and cheap to make but vulnerable to herbivores. Tough ones are expensive but last longer.

Root and Branch

Wood density and leaf toughness aren't the only decisions facing a growing tree. There's also the question of how much of each tissue to invest in. There are three options: leaf, trunk, and roots. All have their advantages. Having lots of roots would provide a firm anchor in the soil and suck up lots of water and minerals; a long trunk and branches add precious height; and the more leaves you have, the more light you can gather and turn into food. So what's the best way for a plant to divide up its resources?

Enquist and Niklas tackled this question theoretically. The key consideration is the way water flows through the plant, from root to stem to leaf. The laws of fluid dynamics rule that, to minimize the resistance in the plant's tubes, the mass of roots and stems should be proportional, although not identical, to one another. As a tree's trunk grows, its roots should also grow at a constant rate. The same theory predicts that the amount of leaves—which control metabolic rate—should vary with the tree's total mass to the power of 3/4. This means that a 100-fold increase in trunk mass brings only a 30-fold increase in leaf mass—small plants have proportionately lots of leaves, whereas big ones have lots of roots and trunk. The relative decline in leaf mass also means that bigger trees have proportionately more of their masses below ground. A plant weighing 100 grams has one-fifth of its mass in its roots; this proportion rises to one-quarter for a plant weighing 10 kilograms. Fortunately for them, Enquist and Niklas did not have to dig up and chop up thousands of plants to test their ideas, because other researchers already had. The ready-made data gave a good fit to the theoretical predictions. Across species from conifers to oaks to mahogany, trees of the same size are investing roughly equal proportions in root, shoot, and leaf.

Forests are still complicated places, and allometry cannot explain everything. Trees of a similar size can vary 100-fold in the mass of their leaves. Some of the trees at San Emilio have massive trunks in which they store water, and so are relatively less leafy than their cloud forest counterparts, for whom the problem is more likely to be how to lose water by pulling it through their bodies and out their leaves. There are also differences that reflect evolutionary history: On average, a conifer has more than twice the weight of leaves of a broad-leafed tree of the same size. Populations also vary, as Gentry found. The number of trees in the original gentrasos varies by a factor of 20, from 50 to 1,000. The rules of networks and energy set the limits of what is possible, but a harsh winter, strong gales, or explosion in the caterpillar population may prevent any one forest from approaching these limits. Like animals, plants have some ability to bend these rules, depending on where natural selection pushes them, and trees specialized to live in shady or sunny, or wet or dry, places will need different strategies. The chemical variation in leaves shows one way that plants can change their biology to compensate for their metabolic limits. But the regularities that emerge on a global scale show that, although their routes may differ, the world's forests have arrived in the same place.

Leaving Santa Rosa, we head south again. Stops include one to buy beer and one for the police to flag me down for speeding. As I wait sheepishly to be given a ticket, the policeman waits for us to offer to "pay the fine here," which as it happens would be rather cheaper. By the time we realize the confusion, he has given up and cheerily sends us on our way, unfined. (I suspect that Brad, who is in the passenger's seat, has used the Jedi mind trick.) The last forest on the trip is back in the clouds, high on the slopes of another volcano, Barva. We collect leaf and wood samples in the general area of a gentraso that Brad surveyed in the past. Hiking out of the woods with our final bags of specimens and our final crop of chigger bites, a male quetzal flies across our path, pursuing a female.

I think of Gentry, who would have marched through this forest putting a Latin name to every tree. Then I think of the message of the gentrasos: All that seems to differ between forests is the number of species that live there. Trees across the world are remarkably constant in the ways in which they use energy and grow, and this uniformity leads to a constancy of structure in the world's forests through space and time. So why is life diverse? If all living things use energy in the same way, why do they come in so many different types? If life is a struggle for energy, why does the cloud forest around me contain scores of tree species, rather than just one that has managed to outdo all the others? If you've seen one tree, why haven't you seen them all?

8   THE CULT OF SANTA ROSALIA

Iₙ THE SPRING OF 1958, G. Evelyn Hutchinson, a Yale University zoologist, was on a field trip in Sicily. A specialist in the biology of lakes, he was looking for a type of water bug known to live on the island. Most of the land was farmed, however, and there were hardly any suitable pools. Hutchinson's luck changed on a sightseeing trip up Monte Pellegrino, a hill outside Palermo. There, in a small pond, he found two species of the bugs he was seeking. The pond was fed by water trickling out of cracks in the mountain, just below a church dedicated to a twelfth-century hermit, Santa Rosalia. The saint was already the patron of Palermo; in his gratitude, Hutchinson added evolutionary biology to her duties.

Yet Hutchinson saw a riddle in his good fortune. Why were there two species of bug in the pond and not one—or 20? He decided to use his forthcoming address to the American Society of Naturalists, of which he had just been elected president, to pose the question. His presidential address, titled "Homage to Santa Rosalia, or Why Are There so Many Kinds of Animals?" is probably the most influential lecture in the history of ecology.

G. Evelyn Hutchinson (1903–1991).
Credit: Yale Peabody Museum, New Haven, Connecticut.

Evelyn Hutchinson made no great discoveries about nature. Nor was he a particularly brilliant fieldworker or experimentalist; in fact, he could be rather butterfingered. Early in his career, Hutchinson pioneered the use of radioactive isotopes to track the movement of chemical elements through living things. Rowing out onto Linsley Pond, in Connecticut, with a vial of radioactive phosphorus, he spilled the contents into the boat. The only way for him to get the phosphorus into the lake was to scupper his craft.

Hutchinson's genius lay rather in his combination of intellectual gifts. Like many biologists, he took a sensual delight in the natural world. In his case it was part of a broader aesthetic, a general fascination with beautiful things. He loved ornament and variety, and late in life wrote a book on decoration in Tibetan and Indian culture. But along with an eye for detail came a gift for spotting patterns—things, he came to believe, were connected. Ecologists often focus their thinking on either the organisms they are studying—counting their numbers, measuring their attributes—or the connections between them, such as the flow of nutrients through a food chain. Hutchinson could see nature's objects and processes simultaneously and so see ecological problems in new ways. This talent, combined with his belief in the power of mathematics to describe natural systems, made him one of the key figures in the transition from natural history to ecology, from cataloguing the living world to explaining it. He was also as much the midwife as the father of modern ecology. Born in Cambridge to a well-to-do academic family, at Yale he was the archetypal Englishman abroad. His exotic demeanor coupled with an exotic erudition attracted a group of remarkable students, many of whom had never thought of becoming ecologists before they met him.

One of the reasons Hutchinson's question has been so stimulating is that it is rather vague. One of the things he meant was: Why is there the number of species that there is and not some other number? He was particularly interested in the idea that, in the absence of environmental upheavals such as floods, droughts, fires, ice ages, or meteor strikes, the number of species in a place arrives at an equilibrium and in whether there is a theory that can predict and explain this number. Fifty years after Hutchinson's lecture, there is no one theory that can do this; or rather, there are lots of them, but all have their caveats and exceptions, and all inspire argument, which at times has been bitter. None of the ecologists following in Hutchinson's wake have quite been able to draw the teeth of his question. To quote one of them, we still do not know why there are 700 species of birds in North America and not seven.

Job Opportunities

In the *Homage*, Hutchinson suggested that one explanation for the number of different species lay in the range of possible biological professions and addresses. Belonging to a species is like having a job: It's a specialized way of making a living, and an evolutionary choice that closes off other employment options. By becoming excellent at one way of life, through adaptation, animals and plants become inept in others. You would no more set a sheep to catch a rabbit than you would employ a plumber to cut your hair. Tropical orchids would struggle on the tundra. Cows are good at digesting grass, bad at ant eating; pangolins, vice versa. Species divide up resources, and each species can exploit some so well that it can monopolize them. But due to life's ubiquitous trade-offs, the ability to hog some resources comes at the cost of being able to use all of them, leaving other jobs vacant. If all organisms use energy in the same way, maybe diversity reflects the number of different ways to get it.

The most fundamental difference in biological jobs is between those organisms that get energy from nonliving sources, called autotrophs, or self-feeders, and those that consume other organisms, called heterotrophs. Plants are the most obvious autotrophs—they use sunlight to build carbon dioxide into sugar—but not the only ones. Some microbes can fuel themselves from the chemical bonds in compounds such as ammonia and methane, using geothermal heat to break them apart. The first autotrophs, born in the ocean more than 3 billion years ago, probably did something similar.

Heterotrophs feed on sunlight by proxy, consuming the cells around them. This is how a food chain works, with autotrophs such as plants feeding herbivores and predators eating the herbivores and each other. The most obvious heterotrophs are impressive beasts such as bison, giraffes, lions, and us, but the vast majority are unobtrusive organisms such as fungi and microbes that get their energy from decomposing dead material. The worms that eat us when we die are above us in the food chain. In fact, ecologists now speak of food webs rather than chains, because in any place most species eat several different kinds of plants, animals, or both and in turn have to avoid several different predators or parasites. Food chains can become loops—the worms might get you in the end, but in the meantime if you eat a chicken that feeds on worms, you can reestablish your trophic superiority. A complex network of links is needed to describe who eats whom.

One way to create diversity is to put another link in the food chain. But as a biological job creation scheme, such a ploy is not very effective. Most food chains are only four or five species long. The ecologist who worked out why they are not longer was a protégé of Hutchinson— Ray Lindemann. Hutchinson and Lindemann were among the first scientists to think about how the patterns shown by a group of different species living in the same place, called an ecological community, could be the result of the way that energy flowed through them—"If the community is an organism, it should be possible to study its metabolism," Hutchinson once wrote. And before he moved to Yale to work with Hutchinson, Lindemann spent five years studying a Minnesota lake called Cedar Creek Bog, plotting the links between plants taking up energy and materials, consumers eating plants and one another, and decomposers feeding on corpses and recycling their nutrients to plants. In the process, Lindemann realized that, in energetic terms, living things are leaky vessels and that the loss of energy as living matter passes up the food chain can explain the pattern of declining numbers and diversity.

For starters, Lindemann reasoned, the laws of thermodynamics prevent energy from being perfectly converted from one form to another. But there are other leaks besides this one. Much of the energy a plant or animal obtains will be used up before it is eaten, and much of it will be converted into tissues that are indigestible to its predator, although not to microbes in the predator's gut or the soil. A large portion of the energy entering one level of a food chain is lost. All flesh is grass, but only about 10 percent of the energy in grass makes it into herbivore flesh. Predators eating herbivores are similarly inefficient. (The efficiency of consumers varies between about 5 percent and 50 percent in different food webs; researchers have sought regularities in the amount of energy passing between levels in a food web, but in this instance the variability of nature has so far confounded them.) So the first predator gets only about 1 percent of the energy in the plants. This is why you can feed more people with a tonne of wheat by baking bread than by feeding the grain to burger-bound beef cattle. The longest food chain found so far has nine members, counting the autotroph as number one. The calories in the top predator have been through eight other bodies, and about 99.999999 percent of the sun's energy has seeped away en route.

In the *Homage*, Hutchinson pointed out that several other things, besides leaking energy, limit the length of food chains. For example, each predator tends to be larger than the one below it. If you want to kill something, it helps to be bigger than it, although this is not true for herbivorous insects and parasites, such as caterpillars, ticks, and mosquitoes, which are important though poorly understood players in food webs. So, in addition to there being less energy at the top of the food chain, the largeness of top predators forces them to spread out. Eventually, each animal needs so much land or water that its lifestyle becomes unsustainable—it will be impossible to find mates, for one thing. This restriction puts a limit on how big and fierce animals can become. Cannibalism further shrinks the potential length of a food chain by reducing the amount of energy available to the next level up. Predators that take their prey from several levels in the food web— omnivores, in other words—have the same effect. A super-efficient predator might eliminate the species between it and those lower down the chain. This is what humans are currently doing in the sea and in the bushmeat trade. As big species such as cod or apes disappear, smaller ones such as herring and rats come onto the menu.

After moving to Yale, Ray Lindemann wrote up these ideas in a paper called "The Trophic-Dynamic Aspect of Ecology." The editors at *Ecology* rejected it as too speculative, but Hutchinson was able to convince them of its worth. It is now considered one of the field's seminal works and helped found the field of ecosystem ecology. But Lindemann never saw it published: While it was in review, he died of liver disease, at age 26.

The Paradox of Diversity

So biodiversity is not a product of long food chains. Most diversity is found within the levels of food webs rather than between them. The 1,500 orchid species in Costa Rica all occupy the same level of the web, as do the thousands of herbivorous insects that feed on them, or the birds and lizards that eat the insects, and the birds of prey and cats that catch the birds. This struck Hutchinson as paradoxical. By the late 1950s, lab and field experiments, on microbes in a jar or grasses in a plot, had already shown ecologists that, if two species were made to fight over the same suite of resources, one would always eliminate the other. Yet diversity is ubiquitous. Seemingly featureless places, such as prairies, invariably contain many species of grasses and herbs. Plankton puzzled Hutchinson most of all. The open water of a lake seems a homogenous environment, like a giant lab flask, yet somehow hundreds of species are able to share it. It takes a lot of effort to minimize the number of species in a place, as gardeners and farmers know.

Besides posing a puzzle, the lab experiments gave a clue as to what might be going on. The outcome of contests depended on the conditions in the arena—tweaking the temperature, or the amount of nitrogen in the soil, changed which species won. This variability seemed to open up more job opportunities for species, more potential specializations. Species can also vary by when they breed, the temperature they prefer, whether they grow best in shady conditions or sunny, whether they are active at night or day, and in many other ways. A few years before his encounter with Santa Rosalia, Hutchinson devised a way of thinking about a species' place in nature—its niche. He saw the niche as a set of coordinates specifying each species' lifestyle choices and how it fit into biological and environmental space. One axis might represent the temperature range a species could tolerate. Another might be the range of food items that species would eat. Tree-dwelling birds could be defined by the height at which they built their nests.

Plants would have a range of light or moisture levels within which they thrived and beyond which they withered—and so on, until all the species' habits were accounted for.

Natural environments vary in space and time. They contain wet and dry spots, leaves and fruit, air and soil. Their temperature, rainfall, and altitude all vary. This variety creates a niche space too broad for any one species to monopolize, and so evolution will favor some degree of specialization. Hutchinson argued that the spread of niches in environments allowed species to coexist, as long as their needs did not overlap too much with those of any another species.

A species' niche is a force field around it, from which it must exclude others or perish. The strength of the field varies, however. Organisms do best in environments close to the center of their niches and worst at the edges. A plant will usually grow quickest at a temperature just cooler than the hottest it will tolerate and will struggle as it approaches its limits; a hawk that prefers to catch mice and voles might be capable of tackling a beetle or a sparrow but will only attempt it if nothing else is on offer. At the edge of a species' niche, other species better adapted to exploit those conditions can muscle in. For example, in Scotland the barnacle *Chthamalus* lives on the higher, drier parts of rocky shores. It has no trouble surviving in the wetter environment lower down, but normally other, faster-growing species crowd it out. But to achieve their fast growth rate, these competitors have sacrificed their ability to withstand drying out at low tide, leaving *Chthamalus* a piece of the beach to itself.

Put another way, evolution favors animals and plants that can avoid competition by moving into less crowded regions of ecological space. As Darwin saw, the struggle for existence brings about natural selection and results in life forms that fit their particular environment. Bats are less agile in the air than swifts and swallows, but they are better at navigating in the dark. By dividing up the day, both groups can feed on flying insects. Giraffes' long necks allow them to reach foliage that other browsers cannot, dividing up food in space. So one reason for there being lots of species is that there are lots of potential niches. This is the multiple Goldilockses view of biodiversity: The three bears' breakfasts can feed more girls if some like either cold or scalding porridge.

Healthy Competition

Another scientist inspired by Hutchinson, Robert MacArthur, developed these ideas about niches. MacArthur was another who thought that the job of ecologists was to spot patterns—even if this meant ignoring some of the detail—to find their causes and to express them in elegant mathematical language. In this way, he wrote, the study of nature could avoid "degenerat[ing] into a tedious set of case histories." He believed that ecologists should be more like physicists and less like historians. They should seek mechanisms to explain what they saw, not just tell a story of how things might have gotten to be that way. More than any other ecologist, he championed the approach to modeling taken by West, Brown, and Enquist—of building theories that make predictions, based on a set of assumptions about underlying causes, and then testing them against data, rather than starting with the data and ending up with a model that describes it. His models were not ways to make detailed predictions. Often their conclusions were very broad, giving answers such as "larger" or "more" rather than "3/4". Instead, they aimed to reveal common mechanisms underpinning biology's many facts.

MacArthur had taken a master's degree in mathematics before coming to work with Hutchinson, but his program was not purely mathematical. He always sought to test his theories against data. This approach would lead to modified theories, which would suggest new experiments, and so on. He was also an expert bird watcher. In his Ph.D. work he looked at five species of warbler, living in coniferous forests, which were thought to be so similar in their feeding habits and behavior that they violated the idea that species could not share identical niches. Through painstaking observation, MacArthur found that each species preferred to feed in different parts of trees. One spent its time around the tree base, another at the treetops. One used outer branches, another inner—evidence that apparently identical species have evolved to coexist by finding many subtle ways to divide up resources. MacArthur also showed that the number of bird species in an area rises steadily with the number of forms of vegetation there, such as grassland, scrub, and tree canopies—in other words, more niches means more species.

MacArthur's influence on ecology was equal to Hutchinson's. Besides being a brilliant and original thinker, he had an aura that made him a prophet for his view of life. He radiated enthusiasm for nature and science and, at least toward the like-minded, was generous with his time, attention, and encouragement. His belief that ecology could be explained and generalized was seductive, and those who spoke with him, particularly young scientists, often remember coming away transformed, feeling that things were suddenly much clearer and that they themselves were suddenly much cleverer.

For a decade beginning in the late 1950s, MacArthur, together with Hutchinson and other colleagues, produced theories seeking to explain many of the key problems of ecology. Many of these theories addressed the ways that species might coexist by dividing up resources (there's a story that the problem of sharing resources had occupied MacArthur since childhood, when he had tried to work out the best way to divide a cake). One of the problems MacArthur pursued using this resources-based approach was why some species are common and some rare. Broadly speaking, a few species are common, but most are rare. Ecologists had sought mathematical descriptions and explanations of the numbers of rare and common species in each place for decades, and they still are. MacArthur built a model that envisaged the environment as a stick (a cake would have been just as good) and then broke the stick into pieces of random length. The length of each piece represented both the numbers of each species and the size of its niche: More common species got longer bits of stick. He then looked to see whether the patterns in the lengths of random stick fragments matched the abundances of real species. The image of the environment as a broken stick does not capture the possibility that species' niches can overlap, that two or more can share the same part of the stick. So MacArthur refined it by trying to work out how similar species could be—how much they overlapped on the environmental stick—without one forcing the other out. Working this out, and combining it with a measure of the full range of resources available, might lead to a prediction of the number of species that could share a place. MacArthur also tried to unite competition and evolution, by working out how the pressure to avoid competition might cause competing species thrown together to evolve differences through time. All in all, MacArthur's ideas became a route map for ecology.

The *Homage* came at the beginning of a golden age in ecology. In the 1960s and 1970s there was optimism that a unified theory of ecology and evolution—that would explain how natural selection acting through ecological laws produced the structure of nature—was close at hand. Hutchinson and MacArthur made competition between species the focus of this research program. The strength of competition between species, and the ways in which it could be avoided, came to be seen as the forces that controlled the number and type of species that could coexist and the evolutionary changes that would occur in similar species sharing a habitat. The hunt was on for a set of rules that governed the way animals and plants divided up resources and environmental conditions—niche space—between themselves.

Body size is crucial here. The extent of an animal's niche, and so the number of species that can pack into any part of ecological space, depends on its body size. Besides trying to work out how an organism's size affects its biology, scientists have spent decades looking at the question the other way around—trying to work out what determines body size. Whatever the determining factors are, they seem to favor smallness. There are mammals weighing everything from 2 grams to 100 million grams, but three-quarters of species weigh less than 1 kilogram. As for birds there are more species of warbler than eagle. There are more small tree species than large; the same is true for fish. This seems to be true however you slice up life: There are more species of insects than mammals, for example.

Hutchinson and MacArthur thought that niches had something to say about this too. To a browsing giraffe, a leaf is less than a snack. But the same leaf might feed one insect species nibbling at its edge and another that burrows in and eats it from within. Other insects could live on the tree's bark, bore into its trunk or roots, or trick the plant into growing around them, creating a gall. Small animals use up less ecological space, and so more of them should pack into that space. The analogy that I used earlier, of packing balls into a crate to describe how trees fit into a forest, could apply here also, except that each ball represents a species, not an individual. More recent ecologists have analyzed this packing using fractal geometry. If you imagine the environment's resources as a fractal, rather than a stick, small species will see that fractal at a higher magnification than large ones. Small species will see, and exploit, all the twists, turns, and branches in resource space invisible to larger species. Large animals perceive the world at a coarser scale and need more space—from feeding ground, to waterhole, to breeding ground, to roost—to get life's jobs done.

On Monte Pellegrino, Hutchinson noticed that the two bug species in the pool were of different sizes. If body size determines how much ecological space a species occupies, differences in body size seem an obvious way to occupy different niches and thus avoid competition. Different-sized organisms need different amounts of space and food, they can get and use different types of food, and they can tolerate different environments. Hutchinson sought a rule to explain size differences in coexisting species. In groups of animal species exploiting the same resource, he noticed, each species is often about twice the weight, or 1.3 (the cube root of 2) times the length, of its nearest neighbor. Hutchinson thought that the critical size difference was between the body parts used to obtain food. If you line up the Galapagos finches studied by Darwin in order of their size, each species has a bill about 1.3 times longer than the next smallest finch. Insects go through several different larval stages before adulthood, and each stage is roughly 1.3 times longer than the next. The idea caught on, and many more examples of what became known as Hutchinson ratios were spotted. In 1977 the ecologists Henry Horn and Robert May noted that in consorts of viols and recorders each instrument, as one moves from treble to tenor or tenor to bass, is about 1.3 times longer than its neighbor. They seem to divide up musical space so that each has a separate job and does not compete with the other group members. And in some sets of iron skillets sold together, they pointed out, each is 1.3 times wider than the next. No one would buy a set of five identical frying pans, but like animals the coexisting skillets have been selected to specialize in handling food items of different sizes. Horn and May suggested that Hutchinson's ratios might apply generally to sets of complementary tools.

Jared Diamond, a friend and colleague of MacArthur, offered another example of how competition between species might control diversity. Diamond studied the distribution of 141 bird species on the 50 islands of the Bismarck Archipelago off the coast of Papua New Guinea. Diamond sought the rules—he called them "assembly rules"—that could explain which species lived on which island. He looked at factors such as body size, the altitude a species lived at, where in the trees it fed, what it ate—such as insects, fruit, or nectar—and how far it would travel (many tropical forest birds will not cross even small stretches of water). Some combinations of species, Diamond discovered, were never found together. For example, two closely related fly-catchers that fed by picking insects off vegetation were never found on the same island. The same went for two similar-sized species of nectar-eating birds. Sometimes, this created a checkerboard pattern. Each island was like a square on a game board—if it was already occupied by one species, it was no longer open to others. Other rules were more complex. Which combinations of species were able to live together could also depend on the size of the island and sometimes on the presence, or absence, of other species. Some species were "supertramps" that could tolerate many different conditions and travel long distances. They did well on small isolated islands, but could not gain a beachhead on islands already containing many more specialist species. Diamond's assembly rules may not have provided a quick and easy route to a general understanding of diversity, but they did—along with many other similar studies—suggest a procedure that, applied in specific places, could explain why a certain number of certain species lived there.

Not everyone agreed with Diamond's assembly rules. In particular, a group at Florida State University led by Daniel Simberloff, another mathematics major turned ecologist, argued that ecologists had been too eager to invoke competition as the cause of patterns in nature. The true way to do science, said Simberloff and his colleagues, was to try and falsify a hypothesis, not find data that supported it. Just because a pattern matched one interpretation didn't mean that a true underlying cause had been discovered. There might be several other mechanisms capable of producing the same pattern—just as a Ptolemaic model of the solar system can predict the planets' movements almost as well as a Copernican one. Ecologists should be testing these alternative explanations against one another.

Similarities and differences are easy to spot but difficult to pin down. If you measure enough things for any species you are bound to find lots of differences, but working out which, if any, are the reasons that they can or cannot live together is another matter. Likewise, measure enough traits and you are almost bound to find some that vary in a ratio similar to 1.3:1. The pebbles on a beach come in many different sizes, but that does not mean that the existence of small pebbles forces others to be big, or vice versa—they are just spread out randomly across a certain range. In 1981, Daniel Simberloff and William Boecklen showed that most examples of Hutchinson ratios were more like pebbles than skillets: a random spread, not a matching set. Similarly, Simberloff's team claimed that many of the patterns Diamond had observed would be reproduced if species had strewn themselves around the Bismarck Archipelago at random. They also pointed out that the bones found on Monte Pellegrino, thought to be those of Santa Rosalia and to have miraculous properties, had turned out to belong to a goat. The MacArthurians, in turn, found flaws in the work from Simberloff's team (or the Tallahassee Mafia, as they were nicknamed) and claimed that Simberloff's models were much less random than they at first appeared.

Now, both sides of the debate sound reasonable. Considering alternative hypotheses is important, as is statistical rigor. Devising meaningful models in the face of messy ecological data is difficult. Some studies of competition wilted under tough scrutiny; others stood firm: Several of Diamond's pairs of bird species are far less likely to be found together than chance alone would predict. Scientists spend much of their time picking holes in what other scientists claim to have found out, but it only occasionally gets personal. So why did the debate between the Tallahassee team and the MacArthurians become, in the words of one journalist at the time, "as acerbic and acrimonious as any that has stirred the combative instincts of academia"? One reason was that MacArthur's approach had already polarized ecologists. His ecology and philosophy inspired devotion and admiration in those on his side, but they could apppear arrogant to those who took a different approach, and MacArthur could be dismissive of his less conceptually ambitious colleagues. He hated being wrong but was not afraid to try and push things along by publishing half-formed ideas or suggestive but inconclusive data. Sometimes, mistakes were later found in these studies, which confirmed him as a charlatan to those who disagreed with his science or resented his fame and influence.

MacArthur saw being interesting and provocative as part of his job; tidying up could come later. A few years after he published it, for example, he decided that his broken-stick model was wrong. Theory would naturally run on ahead of data. It takes days to build a mathematical model but months or years to do a field study. But for MacArthur there was no later: He was diagnosed with kidney cancer in 1971 and died the next year, at age 42. His untimely death left many of his theories only partly formed and tested. It also created a power vacuum. Ecology was a young and growing science, and the generation of researchers following MacArthur was hungry for prestige. Continuing his work was one obvious way to do this; knocking it down was another. The theory of competition, said Simberloff, "[had] caused a generation of ecologists to waste a monumental amount of time." Diamond and his colleagues called the criticisms "silly" and "lacking common sense." All the arguing dissuaded the next generation of ecologists from working on these problems; the discipline shifted back toward experimental studies, and researchers wanting to study the large-scale patterns in nature found it difficult to get funding.

By the early 1980s the preceding decades' optimism had dissolved. Many ecologists despaired that studying competition would bear fruit. The criteria that might explain diversity—the heterogeneity of the environment and the competitive interactions between the many species that share it—are each hugely complicated. In models a small tweak to any one variable can change dramatically the number and type of species that can coexist. Measuring them in the wild is also dauntingly difficult. There were also theoretical reversals. It turned out that *any* differences between species could allow them to coexist, so theorizing about or looking for a minimum difference seems meaningless. This made it seem as if the problem was why there were so few species, not so many. Even in a jar, one species can't eliminate another solely through competition. There must be some other factor causing extinction. This factor could simply be chance, a random fluctuation in the rate of reproduction that tips a population over the point from which it cannot recover. It was at this time that Jim Brown set about looking for some other way to explain diversity besides competition. Other ecologists spoke of their discipline as being "in crisis," "repugnantly complicated," and "in a quagmire."

The Modern Niche

French wine makers have a concept called *terroir*, meaning each vineyard's unique combination of soil, climate, and geography. The idea is heavy with mystique, and the link between *terroir* and a wine's flavor is held to be inexplicable and impossible to unpick. Perhaps biodiversity comes about as the result of some sort of ecological *terroir*, with the number and type of species in a place being due to idiosyncratic local conditions interacting in unfathomably complex ways. Perhaps, although species obviously occupy different niches, the search for generalities is misguided, and the niche is a sterile mathematical concept that cannot be measured in the real world. But many ecologists think not, and for them competition and the niche are alive and well.

Some ideas still focus on how competing species apportion resources. In 1976, David Tilman, then a Ph.D. student at the University of Michigan, found that two species of algae could share Lake Michigan because they differed in their ability to use phosphorus and silicon. One species was especially good at getting phosphorus out of the water and so could thrive in a low-phosphorous environment. But it needed lots of silicon, which the other species was better at obtaining. Each had traded off its competitive abilities in one arena against its competence in another, and, as long as silicon and phosphorus were spread unevenly, both could coexist. Since then, Tilman and his colleagues have developed this idea into a more general theory, particularly applicable to plants, which argues that each species is a specialist for a certain set of conditions. Every point in niche space becomes a potential niche—again, the potential number of coexisting species is limitless. Species thrive in some places and wither in others, but because environments are variable many species can share a place.

Another, more recent, idea of how species divide up resources, devised by Mark Ritchie and Han Olff, resuscitates the MacArthurian idea that biodiversity is a question of how many species you can pack into ecological space and combines it with our old friends scaling and fractals.

Imagine a group of species all fighting over a pot of resources, such as grasses for light, water, and nutrients, or grazers competing for grasses, big cats for grazers, and parasites for hosts. What each species needs is not spread evenly across the world. It comes in clumps, such as a moist, well-lit piece of bare ground, or some juicy young shoots, or a carcass. To survive, organisms must find these patches amid the barren areas that cannot feed them—grass seeds must disperse over areas of woodland; wildebeest must follow the rains; lions must find the wildebeest. Frogs must find ponds; blackbirds, worms; fleas, dogs.

Just as not all places are equally hospitable, neither are all patches. They will vary in size and quality. A grass seed could find itself on a prairie or in a crack between some paving stones; a flea could find itself on a glossy young pup or a mangy old mutt. Whether a patch of habitat can keep an animal or a plant going depends on its size and quality. It also depends on the properties of the organism seeking to exploit that patch—principally its body size, because this property controls how much food an animal needs and how long it can go without starving. Big animals need more food, but they can go for longer periods between meals, and so can use large but thinly spread patches. Smaller ones need less food but in higher concentrations. Such a multiplicity of factors makes working out how the environment might control the diversity of organisms seem horribly complicated. A glance at any map shows that each landscape is a unique mishmash of wet, dry, rocky, fertile, high, low, windy, and sheltered places. How can we hope to predict what sort of patches it contains and what sort of life these patches can support? We can look for regularity hidden amid the chaos.

Take a map and scan it into a computer. Get some software to look for all the patches of one sort of habitat, such as woodland, and work out how big each patch is. Then get the computer to divide the patches into size classes, such as 0.1 to 1 hectares, 1.1 to 10 hectares, and 10.1 to 100 hectares, and plot how many patches fall into each size class. The spread of patches sizes turns out to follow a power law distribution. Woods—or grasslands, or ponds—are like earthquakes. There are a lot of small ones and a few big ones, and their commonness declines as their size increases in a regular, logarithmic way. This is why maps and landscape photos need scale bars: Their features look more or less the same at all magnifications. The same goes for patch quality—there are lots of places that offer slim pickings and a few gold mines. So although we can't predict where habitat or resources will be in a landscape, we can predict that they will be divided up like a fractal. In a way this is unsurprising, because the physical processes that create landscapes, such as earthquakes, erosion, and landslides, are also fractal.

Fractal thinking also offers a worm's- (blackbird's-, wildebeest's-, grass seed's- . . .) eye view of natural resources. One feature of fractals is that the finer the scale you use to measure them, the longer they appear. Small species have short measuring devices and will perceive more resources. A forest pool is a drinking fountain to a passing bear, a home to a beaver, and a universe to a single-celled alga. The same scale-dependent view applies to food within patches of resources. The browsing giraffe bites off the whole leaf, getting lots of tough and indigestible plant tissue in the process. The herbivorous insect that tunnels between the leaf's veins eats only the nutritious green flesh, leaving the skeleton behind. How much we see in nature depends on how closely we look at it, which depends on our size.

So small species will find a high concentration of resources, but they need to, because they can only exploit small patches. Larger species will find food more thinly spread but will be able to cover more ground to get it. Goldilocks redux—for a species of any size, some patches will be too small, others will be big enough but too sparsely provisioned, and a few will be just right. The landscape is a template that is filled in with the species fit to each patch of resources.

So if you know the amount of resources in a place, you can predict the fractal way that these resources will be divided, how many species that place will be able to support, and also how different those species should be, because their sizes ought to match the holes in the environmental template. What's most impressive about Ritchie and Olff's idea is that it predicts the number of species that a place should be able to support. The duo tested their model by trying to predict how many herbivore species would be found in each of 28 different East African wildlife reserves (where the quantity of resources, in the form of plant food, can be predicted from the amount of rainfall) and how many grass species should be found in experimental plots on the Minnesota prairie (where the amount of nitrogen in the soil is a good measure of total resources). The results were impressive: The number of species at each site, and their size differences, matched the model's predictions well. But Ritchie and Olff's idea is not all-embracing. It works well only on smallish scales, areas up to a few thousand times larger than the individual home ranges of the organisms under study. For soil bacteria this would be a few cubic centimeters; for a plant it's a field; for a grazing herbivore it's the area of a game park. Over greater areas other factors, such as whether organisms can get between distant patches, come into play and so its predictions start to break down—it can't tell you why there are 700 species of birds in North America and not 7.

Both Tilman's and Ritchie and Olff's theories are based on the Hutchinsonian idea that biodiversity is the result of some sort of balance in how species divide up the environment. But other ecologists think that nature is not in balance and that it is the very disturbances that Hutchinson ignored—the droughts, floods, and fires—that allow species to live together. Even if two species share identical niches, one does not instantly eliminate the other. The speed with which this happens depends on what else is going on in the environment. In a stable environment, competition will run quickly. Nothing else is going on. But in the real world, lots of things can send the competitors back to the starting line, shift the balance between them, or so preoccupy them that they never get around to competing at all. A tree cannot crowd out its neighbor if it is struck by lightning or blown over. But the next lightning bolt will strike another species of tree, allowing a third species to survive. On a human timescale, it looks as if the trees in a rain forest have struck a happy balance. In fact, they might be strangling one another, but so slowly and haphazardly that we never see a killer blow. Taking competitors out of the game reduces the strength of competition, perhaps to a level where competition is not strong enough to curb diversity. Maybe the harshness of the world prevents nature's bus from ever becoming so full that species must fight over seats.

Another thing, besides environmental hazards, that opens up seats on the bus is species' habit of eating one another. This could be a particularly effective way of maintaining biodiversity because, whereas storms strike all the trees in a forest more or less equally, predators and herbivores tend to focus disproportionately on the most common species in their environment. By reducing the numbers of species they feed on, these consumers prevent their prey from overwhelming their neighbors. If starfish are taken off a beach, the mussels they feed on take over, and the total number of species goes down. Likewise, herbivores tend to choose plants that are juicy and fast growing but not well defended. Without the browsers and grazers, these plants would take over. But with them other plants that are slower growing, because they have invested in poisons, thorns, or other defenses, can survive. One idea argues that every tree in the forest inadvertently attracts herbivores that specialize in eating it. Mature trees can cope with this, but it's no place to raise a child—any young tree of the same species trying to establish itself in the same neighborhood is in trouble. These herbivores, however, would ignore nearby trees of other species, allowing a patchwork of diversity to build up. There is some evidence that tropical trees do indeed find it harder to germinate and survive close to members of their own species. The crucial common aspect of these predator effects is that they weigh heaviest on common species. This creates a benefit to being rare that maintains diversity by reducing strong competitors' advantage.

These are just a few of the many ideas, none of them mutually exclusive, that different ecologists currently favor to explain why species do and don't live together. There are variants of these, but they are all twists on the basic concepts of competition and how species avoid it, either by dividing up space, time, and resources, or because of forces beyond their control.

The Antichrist of Ecology

But maybe competition between species is not strong enough to control biodiversity. Maybe species are not adapted to slot into a niche, dividing and ruling ecological space. Maybe they just drift around, with chance and history controlling the combination found at any one place and time. Maybe ponds, or forests, or coral reefs are in flux, not in balance. Maybe there had been three species of water bugs living on Monte Pellegrino the week before Hutchinson showed up, but one had died out. Or maybe there had been one, and a pregnant female of another species had only just flown in, and another species arrived the week after he left.

Nature certainly changes. Deposits of tree pollen in Europe and America show that over the past few thousand years species have come and gone and that a place can contain a different set of plants from one millennium to the next. And for the past few millennia, humans have been inadvertently challenging the idea that species are packed tightly into ecological space. In many places the dominant wild plants and animals are recent colonists, aliens that travel along with people. In the past two centuries the native prairies of California have been almost entirely replaced by immigrant European grasses—the state is now home to more than 500 introduced plant species. If the natives were wedged firmly into their niches, we would expect invaders to have difficulty establishing themselves. Perhaps immigrants could only succeed by being bigger or meaner than the residents, like the carnivorous snail *Euglandia*. This species was introduced onto the Pacific island of Moorea in an attempt to control another exotic, an African land snail, but instead ate its way through the native tree snails to the extent that several species went extinct and others now survive only in captivity. If all invaders were like *Euglandia*, this would challenge the idea that diversity has reached some stable point, but it would at least preserve the notion that environments were full and that niches must be fought over. But, while invasive species are rightly thought to be a serious conservation problem, many, perhaps most, exotics slot into their new homes without causing a fuss and with no discernible influence on the organisms already there. One survey of alien species in estuaries estimated that more than 90 percent of immigrants had no discernable effect on the residents. If environments were running at full capacity, we would not expect this. And we would expect places with more species to be harder to invade. This too seems not to be the case.

So ecological communities are fluid, unstable things. But just because there might be no balance of nature does not mean there can be no laws of nature. A whole class of theories, a school with a history just as long as that to which Hutchinson and MacArthur belonged, offers alternative explanations for biodiversity based on the idea that nature is unstable. The difference between these and niche-based models is that they rely on probabilities, rather than causes, and that instead of competition they see migration as the most important ecological process.

One of the boldest and most controversial current ideas about biodiversity was born in the same place that Enquist now works, the San Emilio plot in Guanacaste, Costa Rica. This plot was established in 1976 by Steve Hubbell. Hubbell came to Costa Rica to study the ants and bees, to track their movements and work out how they fed. One of his main interests was in different species' niches. He had already discovered how different bee species drank flowers' nectar at different times, some arriving en masse when the blooms were fresh, others coming along later in ones and twos to mop up the leftovers.

To track the ants' movements, Hubbell began marking the trees. Once he had marked all the trees in his study plot, he was struck by how patchily the different species were arranged. Many species lived closer to others of their own kind than would be predicted if the trees were driving their offspring away by attracting specialist herbivores. When Hubbell looked at the species in San Emilio, the individuals of most were either spread at random or found in groups, not evenly spaced.

Hubbell began to suspect that the patchwork of tree species in his forest was too irregular to be explained by patterns of herbivores or the range of niches found there. In a mature forest, vacancies appear unpredictably. When a tree falls, it creates a gap that the seeds and saplings that have wound up in that spot race to fill. It seemed to Hubbell that who would win the race was as unpredictable as where the gaps would be, and that being in the right place at the right time was more important than being specialized to a particular set of conditions. Hubbell lost interest in ants and bees and began thinking about how the forest could have come to be.

Forests are big. Trees are long lived. So to understand forests, you

need to study a large area over a long time. Al Gentry's rapid-fire cen-
suses had shown that it was possible to collect detailed yet large-scale
information on forests. Hubbell took things further. In 1980 he and
his colleagues set about mapping and identifying every tree in a 50-
hectare block of forest—equivalent to 500 gentrasos, nearly double the
combined area of all of Gentry's original plots—on Barro Colorado
Island in the Panama Canal, where the Smithsonian Institution has a
research station. Unlike Gentry's surveys, this one would be permanent. You could return year after year and see who had died, who had
arrived, and who had grown and by how much. This method, called
the Forest Dynamics Plot, has now become the standard for long-term
studies of forests. There are plots in forests from Puerto Rico to
Cameroon to Thailand; in total 3 million trees belonging to 6,000 species
are being monitored.

Over the years, with each census, Hubbell became more convinced
that the spread of niches was not controlling the diversity and abundance of the trees on Barro Colorado Island. Most species didn't seem
to care whether they grew in a small or large gap or in a shady or well-
lit spot. Each individual was surrounded by a diverse and unpredictable
set of neighbors. This unpredictability ought to have prevented trees
from evolving into specialized competitors, because they could not
predict what they would be competing with. Species also came and
went. Four shrubs went extinct in the dry El Niño year of 1983. Others,
confined to just one corner of the island, seemed to be recent arrivals.

When not counting and measuring trees, Hubbell spent his time
playing with theoretical models, trying to find the levels of birth, death,
and migration that would reproduce the number of species, and their
population sizes, that he saw on Barro Colorado. Fortunately, there
was a ready-made theory at hand, and, perhaps paradoxically, it too
was the work of Robert MacArthur.

In 1963, MacArthur, working with Edward (E. O.) Wilson came up
with an idea to explain the number of species on an island. MacArthur
and Wilson's model could hardly be simpler. The number of species on
an island, they reckoned, is a balance between the rate at which new
ones turn up and the rate at which the inhabitants go extinct. In its
most basic form this model doesn't need niches, or competition, or

any sort of interaction between species. It's just a question of probabilities. On an island with few species, almost every immigrant is likely to represent a new species. But if an island has lots of species already, the chances that an arrival will be a novelty are small. So the immigration rate of new species falls as the number of species on an island rises. Each species also has a certain chance of going extinct. The more species you have, the more will go extinct in any time period. The extinction rate rises with the number of species. The island's total number of species will be where these two trends balance out. The number of species will be stable, but which species they are will change.

Of course, the real world is unlikely to be that basic, and many ecologists have customized MacArthur and Wilson's model to try and make it more realistic. They have looked at how competition might reduce the chances of immigrants establishing on a crowded island or how the distance from the mainland affects the immigration rate. But to get started Hubbell needed just the basic model.

So, Hubbell thought, perhaps the tree species on Barro Colorado were just a random subset of the species in the wider region, the whole of Central America. New species could arrive blown on the wind, or stuck to a bird's foot, or fall to the ground in monkey droppings. Sometimes a storm would fell the last tree of its kind on the island, making that species locally extinct. But no species is better suited to life on Barro Colorado than any other. In the 1980s, Hubbell began tinkering with the chassis of island biogeography—keeping a low profile because of the field's recently bloody history—to see if he could get it to reproduce the pattern he saw in Panama.

For starters, Hubbell stopped looking at the rate that *species* arrived in and left the forest and switched the emphasis to the birth, immigration, and death of *individuals*. Then, instead of starting with an empty island, he assumed the opposite—that the forest was saturated. There were no gaps in the canopy, and for a new tree to grow, an old one must go. His next tweak was to add new species by evolution, as well as migration, in the form of a number giving the probability that any individual will give birth to a new species. Although this scenario is unlikely—species are far more likely to arrive in a forest from outside than by evolving there—plants can sometimes produce offspring that are genetically very different from themselves. Plus it was the easiest way to add evolution to the model without complicating the mathematics unduly.

What Hubbell never included was specialization. He wanted to see how much he could explain by imagining that forests are filled with an average universal tree. If real trees were close to this average, with broad niches that overlapped with the others in the forest, his theory was potentially powerful. If they were very different, his ideas would probably not explain anything.

In the resulting model the species that fills the gap depends not on how wet, or shady, or rich in nutrients the gap is. It just depends on who gets there first. The probability of surviving or reproducing is also equal for all species. The reason for there being so many different types of life is that the differences don't matter.

Hubbell's eureka moment came at 2 a.m. one night. He realized that by multiplying the (very small) number giving the probability that a new species would spring into being with the (very large) number of individuals in the whole region, such as Central America, he got a new number that seemed to control the entire community. If you fed in the right value, it predicted the number of species in an area. It also predicted relative abundance: which species should be common, which rare, and how common or rare each should be. Hubbell seemed to have cracked ecology's two great problems at once.

He called his magic multiple the universal biodiversity number. When this number is small (when the population is small or the probability of speciation low), you get a forest dominated by a few species, similar to the conifer forests of Canada and Siberia. When the number is large, the forest has many species, and the numbers are more equal, as in a tropical forest. In the universal biodiversity number, Hubbell thought he had discovered the $e = mc^2$ for ecology.

In Hubbell's model the population size of each species fluctuates at random. In effect, this means that all species eventually go extinct in a place. This is so because no population can become infinitely large, so in the long term the only way for any species to go is down, even if it originally dominated the entire forest. Eventually it will have a run of bad luck that wipes it out. But, particularly for common species, this extinction will take so long that to a human observer—particularly one on a three-year research grant—the forest will seem like a stable set of species. Hubbell called his theoretical framework neutral ecology, a deliberate echo of an older theory called neutral evolution, which says that much of the change in an organism's DNA confers no selective advantage or hindrance on it.

Neutral ecology is not biodiversity's theory of everything. It can only predict the species within one level of a food chain; for any one place the plants, herbivores, carnivores, and decomposers would each need their own model. Neutral models are also likely to be more successful at predicting the diversity of plants than animals. Forests are planes tessellated with individuals, a lot like Hubbell's theoretical environment. Species share similar requirements, new individuals can only establish themselves when a gap opens, and once they have settled they cannot move somewhere more favorable. Animals are much more flexible in both what they consume and where they consume it. To a rain forest bird, the world is not a two-dimensional plane—it's a complex three-dimensional space. In many ways, animals can carve out their own ecological spaces.

But on its own terms, neutral ecology is remarkably powerful. Get the right population size and the right balance between birth, death, and speciation, and a near-perfect re-creation of a forest pops out of the model. With just three numbers, for example, Hubbell was able to predict both that a Malaysian rain forest should contain 800 species (which it did) and the individual abundances of each species.

Neutral evolution was originally controversial but is now well established. Neutral ecology is still in its controversial phase. About half a dozen journals rejected Hubbell's first paper describing the universal biodiversity number. It eventually sneaked out in 1997 in an obscure journal called *Coral Reefs*. And Hubbell's model still makes many ecologists deeply uncomfortable. At one conference, someone called him "the Antichrist of ecology." This is partly philosophical. Biologists have traditionally studied what makes species different from one another, and they are impressed by how many aspects of the species they study seem to fit a particular environmental job. Neutral ecology seems to start from assumptions that are clearly wrong. It seems like throwing ecology away.

There have also been challenges to the theory's science. In particular, several teams have claimed that there is more structure in nature than neutral ecology can account for, that who lives with who *is* predictable, suggesting that the species in a place are evolved to match the conditions there and are not a random subset of the whole. Neutral ecology also seems best at predicting patterns in diversity at intermediate scales, from a few hundred square meters to tens of square kilometers. On larger scales, environmental differences become important—just as a plant from the tundra would not survive in a desert. On smaller scales, interactions between individuals do seem to have an effect on who can live with whom, and ecologically similar species are less likely to be found very close together.

Experiments can test whether plants live where they do through chance or because they have found their niche. You just move the plant from its wild home to somewhere else. In a world of niches, residents will be better off and migrants will struggle. In a neutral world, travel will be easier. A team led by Graham Bell, an ecologist who developed models of neutral ecology independently of Hubbell, found that plants were quite happy to be moved around Canadian forests. There was no sign that they had a strong preference for a particular environment or set of neighbors. And neutral ecology makes predictions about time as well as space: Widespread and abundant species are more likely to be old, because it takes a long time to become common relying on dumb luck alone. Rare species are more likely to have evolved recently. Comparing DNA, which gives clues about how long ago a species came into being, might answer this question. A bit like Kleiber's rule for body size and metabolic rate, the real world is bound to deviate from neutral ecology's predictions: how much it deviates, and how, will show what needs to be explained by other means, such as competition, and what doesn't.

Patterns or Explanations?

There is also the issue of whether a mathematical model that reproduces a natural pattern, however uncanny its accuracy, really explains what is causing that pattern. Tweak the universal biodiversity number just right and it will give a good fit to patterns of biodiversity—but does it really capture what, if anything, is controlling which species live where? Some ecologists have claimed that other mathematical models of commonness and rarity give a better description of nature than Hubbell's neutral scheme. And models based around niches and competition, if they are detailed enough, can do just as good a job of producing patterns in diversity as neutral models.

Hubbell has an interesting take on this issue. He draws parallels between neutral ecology and physics. Physicists, he points out, tend not to worry about the underlying mechanism of their theories as long as they can describe nature. No physicist thinks Newton's laws wrong because they do not describe how gravity works. In the same way, the laws of thermodynamics or electrical resistance were discovered by observation and correlation, not through studying the properties of individual atoms or electrons. Like Jim Brown, Hubbell evokes the physics that uses statistics to describe the behavior of groups. The laws that describe the properties of gases were not discovered from the bottom up, by considering what every molecule was doing. They are exercises in emergence, describing how the group as a whole behaves, based on an assumption that all molecules are identical and respond to their environment identically. Ecological problems are harder than physical ones because their particles—living things—are so much more variable and unpredictable. But Hubbell and Brown believe that ecology can adopt some aspects of physics' worldview, by looking for statistical regularities in the behavior of nature as a whole, rather than trying to assemble it from its parts. To look for a deeper level of explanation is to step onto a slippery slope: There is always something more fundamental waiting to be explained, another term to be bolted on to the equation. The best criteria for a good theory, Hubbell says, are that it should be simple and make accurate predictions—"that it should fail in interesting ways"—not that it should be an exact replica of the thing it is trying to explain.

There are certainly aesthetic criteria to what makes a good theory, the main one being elegance through simplicity. Making a model more complex will always make it reproduce patterns more accurately, but it does not necessarily make it a better tool for understanding nature. The more parameters you have, the more things you must measure to test them. The mathematics is also wont to become opaque—it can be as hard to work out why a very complex model gives the predictions it does as it is to understand the system you are trying to model. On the other hand, neutral ecology is not like the physical theories Hubbell invokes, because it says definite things about how the patterns it reproduces come about. It's true that the equations relating the pressure of a gas to its temperature cover up a mass of details, but these equations are only about phenomena, not mechanisms. They are formal descriptions of the system that go from data to theory and back again, like allometry. Neutral ecology, on the other hand, *does* contain ideas about mechanisms, even if they are very hard to get a handle on. Like a good MacArthurian model, it goes from theory to data. This makes the model potentially disprovable, which is a good thing. The rate at which new species appear, and the total number of trees in Southeast Asia, either is what Hubbell's model predicts or it isn't.

The exciting thing about neutral ecology is that it is challenging and counterintuitive, and its implications for our view of nature are a powerful incentive to carry out the studies that could test it. Herculean efforts such as FLUXNET and the Forest Dynamics Plots show that such tests, although daunting, are possible. This will all take time. Neutral evolution took more than a decade to become accepted, and nobody needed to measure 3 million trees. It will take at least as long to put neutral ecology through the wringer.

Unraveling the Web

A theory of biodiversity ought to be able to predict the consequences of changing biodiversity. Achieving this goal would be more than a matter of intellectual satisfaction. Plants, animals, microbes, and fungi keep soil fertile and air and water clean; they pollinate crops and control their pests and provide food, fuel, building materials, and drugs. The services provided by wildlife are estimated to be worth $33 trillion to humanity each year. Lately, humans have been unpicking this life-sustaining fabric at a rate seen rarely in Earth's history. The fossil record shows that about 10 species go extinct each year, but currently about 30,000 species are lost each year (both numbers are very approximate, and the real value could be different by a factor of 10). We are in the midst of the sixth mass extinction of the past 600 million years. At the current rate, between a quarter and a third of all species will have disappeared within 50 years. We have also been inserting species into new places, taking plants and animals around the world deliberately, as crops, livestock, pets, and ornaments and, inadvertently, as stowaways in cargo and ballast water. What will result from all this messing with biodiversity?

It might seem that the lesson of neutral ecology, and of the Enquist group's work on the structure of forests, is that, if many of the patterns in nature are not caused by the differences between species, then losing or gaining a few isn't going to matter much. On the broadest scale, this might be true (although we are losing more than a few species). Most invaders have little effect, and most species are probably loosely packed into their surrounding environments. Pulling them out, or cramming a new species in, does not have much impact on the species around them. But on a local scale the details of species matter. Ecosystems are complex and unpredictable and do not equal the sum of their parts. There are many examples of the removal or addition of a single species having serious consequences. And until we make the change, we can hardly predict what these influential species will be.

Often, the most dramatic effects of changing biodiversity come from chain reactions acting through food webs. For example, as D'Arcy Thompson predicted, by the end of the nineteenth century Russian hunters seeking fur had nearly driven Alaskan sea otters extinct. What he couldn't have foreseen is that, because otters eat sea urchins, the urchins boomed in their absence. Urchins eat kelp, so the crash in otters led to a crash in kelp. Kelp helps stabilize shorelines, so storms and erosion became more damaging. Otters recovered, but twentieth-century overfishing may have triggered the same chain reaction, by removing the food of killer whales and forcing them to eat more otters.

Species like the sea otter, which directly or indirectly have a large effect on many of the other species in their environment, are called keystone species. They need not be predators: They could equally be plants or herbivores. It is hard, if not impossible, to tell from looking at a food web which, if any, species are keystones—and for most places we have no idea of the structure of the food web. Ecosystems are a bit like the game of Kerplunk!, that staple of 1970s childhood, in which the aim is to gradually dismantle a lattice of plastic sticks without letting the marbles the lattice supports fall through. It is seldom obvious which sticks it is safe to remove and which will trigger an avalanche of marbles. We are playing Kerplunk! with the planet.

In addition to each species' ecological influence, the sheer number of species in a place is important. One rarely, if ever, sees a tropical tree stripped bare by caterpillars. On the other hand, places where there are fewer species, such as a field of cabbages or a conifer forest, are much more prone to plagues of pests, such as the spruce bark moth, which has killed tens of millions of Alaskan trees in the past 20 years. Experiments likewise show that places with more species tend to have a greater biomass and are less prone to booms and busts in their populations.

There are several reasons this might be so. Diverse ecosystems could have some redundancy—a more diverse portfolio—so if one species disappears, there is another to do the same job. And a diverse group of species could spread life's load more evenly, with each responding differently to environmental conditions and coming with its own set of predators and diseases. Everyone suffers a bit—on inspection, many of the leaves in the Costa Rican cloud forest are riddled with holes chewed by insects—but disasters become rare.

Biodiversity also makes itself felt through its influence on the structure of food webs. If a predator is too effective and too specialized, it will cause a crash in its prey's numbers, followed by a crash in its own numbers as it starves. But adding a competing predator that feeds on the same prey weakens this interaction, because some of the energy in the prey species takes another route through the food web. In computer models of food webs of this sort, the dynamics of the community as a whole become more stable as more species are added—the total number of organisms, the biomass, and the chemical processes all become less variable and quicker to recover from perturbation. Each individual species, however, becomes more likely to go extinct because it has a smaller population.

A Unified Theory?

It has been a long time since I mentioned metabolic rate. Because metabolism controls processes in individuals such as growth and life span, some ecological properties, such as a population's density and growth rate, seem closely linked to metabolic rate. But how metabolic rate might be linked to the quality of life—biodiversity—as well as its quantity is still unclear. But food webs, which describe the way that energy is split between species, suggest a link between the way that individuals use energy and biodiversity.

Like blood vessels, food webs are networks through which energy flows. This is one of the things Hutchinson meant when he commented on studying the ecosystem's metabolism. But we cannot carry the analogy too far. The two types of networks have rather different geometries, and food webs are dynamic—their links grow weaker and stronger, switch on and off, and rearrange themselves in a way that blood vessels do not. But the way that energy flows in and out of the environment, via living things, depends fundamentally on the flow through individuals. And, as we have seen, the flow of energy through individuals—their metabolism—depends on their size, temperature, and chemical supplies and demands. So it is reasonable to expect that size and metabolic rate will have something to say about food webs and ultimately diversity.

There are also hints that links are beginning to form between niches and metabolic ecology. In a 2003 book, *Ecological Niches: Linking Classical and Contemporary Approaches*, designed partly as a response to the challenge of neutral ecology, Jonathan Chase and Matthew Leibold describe how ideas about niches are changing: "Rather than use vague concepts such as 'niche overlap' and 'niche breadth,'" they say, "we focus on measurable aspects of the biology of organisms such as growth rates, consumption rates, and death rates." These, of course, are the very things that an understanding of metabolic rate can help to explain.

The energy in sunlight, refracted through the prism of the environment, produces the spectrum of biodiversity. The structure of that prism is fearsomely complicated. It consists of the physical world—the lay of the land, the spread of water, the chemistry of the soil, the changing of the weather, and the other organisms that all living things meet with as food, predators, cooperators, or competitors. Such complexity has fostered intellectual diversity. The tightly focused worldview of ecologists in the 1960s and 1970s has dissolved into a babel of ideas. It can seem as if there are as many theories as there are theorists.

This problem may well reflect the real world: The different things that biologists have suggested might control biodiversity are not mutually exclusive, and it seems fanciful to imagine that they will collapse into some theory of everything. Plants, animals, and microbes each have very different needs and different ways of fulfilling those needs. The reason there were two species of water bug in the pool on Monte Pellegrino in 1958 is surely different than the reason that there were 60-odd species of tree in the Savegre gentraso in 2005, or 59 species of butterfly in Britain, or about 4,300 mammal species on Earth. Diversity looks different depending on the scale on which you consider it, and each scale will probably need its own explanation. But it is possible to imagine a set of ideas, including the processes of metabolism, the ubiquity of power laws and self-similarity, the structures of niches and food webs, and the processes described by neutral ecology that together will answer Hutchinson's teasing question: Why is there biodiversity?

9 HUMBOLDT'S GIFTS

O NE MORNING A FEW WEEKS after my return from Costa Rica, I caught the bus to Oxleas Wood, one of the few remaining scraps in London of the forest that once covered southern England. It was a weekday morning, so thankfully there were only a few joggers and dog walkers about to see what I did next. I stretched out my arms—it's almost exactly a meter from the tip of my nose to the tip of my outstretched fingers—and, taking aim at a likely looking patch of trees, paced out 50 meters, thus completing one-tenth of an amateur's gentraso.

On my way I bumped into 22 stems broad enough to be measured—not many less than you would find in a cloud forest. But 17 of them were oaks. Of the remainder, there was one each of silver birch, holly, hawthorn, sweet chestnut, and sycamore. The last two of these are not even native: The first was introduced by the Romans, the second from France in the middle ages. Strolling around, I guessed that another nine lines would have added beech, ash, rowan, and hazel to my species list, but I would have been lucky to get much beyond low double figures.

I would have gotten much the same result had I compared any European woodland with any patch of Costa Rican forest. Life is more varied in the tropics than in temperate regions and more varied in temperate regions than at the poles. This is one of the earth's most obvious features.

The possible causes are simple. Either species are more likely to evolve in the tropics or they are less likely to go extinct—or both. But working out what might speed up the birth of species or slow down their death has proved much more complicated. Two centuries after the first hypothesis was proposed, there is still no accepted explanation for what causes biodiversity to peak in the tropics. Besides being the oldest question in ecology, it is one of the toughest.

Discovering Diversity

The greater variety of tropical life must have been obvious to European travelers from the start. Science began to come to grips with tropical diversity in the eighteenth century, when expeditions added scientific inquiry to their missions of discovery and conquest. The model was the voyages of James Cook, whose ships observed, measured, and collected everywhere they went. In the process Joseph Banks, the naturalist on Cook's first voyage, acquired the world's largest herbarium.

Of course, besides containing more species, the tropics have a very different climate than the temperate regions. Natural philosophers immediately fastened on to this feature as the cause of tropical regions' lushness. German botanist Carl Ludwig Willdenow discussed this issue in his book, *The Principles of Botany and of Vegetable Physiology*, published in 1792 and translated into English in 1805. Willdenow noted that warmer places had more plant species. He also gave a detailed measurement—it might be the original piece of macroecology—of what is now called the latitudinal gradient in diversity:

> The Florae of different parts of the globe, with which botanists have favoured us, show indeed that vegetation increases with the degree of warmth. In Southern Georgia, according to credible accounts, only 2 wild growing plants are found; in Spitzbergen, 30; in Lapland, 534; in Iceland, 553; in Sweden, 1,299; in Brandenburg, 2,000; in Piedmont, 2,800; on the coast of Coromandel [the region of southeast India around Madras], about 4,000; in Jamaica as many; and in Madagascar nearly 5,000.

Current estimates stand at 25 plant species for the South Atlantic island of South Georgia and 10,000 to 12,000 for Madagascar.

Willdenow did not suggest any explanations of why warmer temperatures led to more diverse plant life. But one of his students, an aristocratic teenager who went on to be one of the greatest scientists of the era, did.

In his old age, Baron Friedrich Heinrich Alexander von Humboldt tried to destroy all records of his childhood, to prevent future biographers uncovering anything unflattering and to prevent any revelations about his personal life obscuring history's view of his science. No matter: Alexander von Humboldt's scientific achievements alone are more than enough to chew on. As a biologist he demonstrated the influence of electricity on nerve and muscle tissue and showed how the electric eel has its shocking effect. As a natural historian, he collected 60,000 specimens and described 3,500 new species. As an anthropologist, he was the first European to study the Incan, Aztec, and Mayan civilizations and deciphered the Aztec calendar. As a geologist he was the first to note that volcanoes come in chains and suggested that this had to do with lines of weakness in the earth's crust. As an oceanographer, despite never sailing on the Pacific, he predicted the existence of that ocean's Peru Current (sometimes called the Humboldt current) from observations of the South American climate. His friends included Goethe, Thomas Jefferson, and Tsar Nicholas I of Russia. He founded the first international scientific organization, to pool and share data on the study of the earth's magnetic field. His last work was a book, titled *Cosmos*, which attempted to describe and unite all scientific knowledge. It was unfinished on his death in 1859, at age 90. The German government gave him a state funeral.

Born into a family of slightly down-at-the-heel Prussian nobility, Humboldt followed his time under Willdenow with study at Göttingen University, then Germany's leading scientific school. After this he studied geology at the Freiberg School of Mines. In 1792 he took up a job as an inspector in the Prussian Department of Mines. In 1796 his mother died and left him a large fortune. Humboldt could now pursue his greatest ambition—to travel. As a child he had read of Cook's voyages and dreamed of emulating them. At Göttingen, Georg Forster, the naturalist on Cook's second voyage, became his mentor.

Quitting the Department of Mines, Humboldt spent the next winter in the Alps, surveying the land, taking weather readings, and practicing to be an explorer. His first plan was to join a British expedition to Egypt, but Napoleon's invasion of Egypt killed the project. So Humboldt went to Paris, then the center of the intellectual world. There he met scientists such as the zoologist Jean-Baptiste Lamarck, the anatomist Georges Cuvier, and Lavoisier's former colleague, the mathematician Pierre Simon Laplace. Humboldt made such a favorable impression on the grandees of French science that he won a place on a planned expedition to the Pacific, a five-year journey over land and sea. Napoleon, however, intervened again, canceling the expedition and diverting its funds into his military campaigns. But if Humboldt had no voyage, he did find a traveling companion, in the shape of the abortive expedition's botanist. Aimé Bonpland, a Frenchman born near Bordeaux in 1773, had qualified as a physician but had little interest in medicine. His true passion was plants, particularly roses.

Humboldt was stiff and formal; Bonpland was a charmer. Humboldt was meticulous and hard working; Bonpland made up his life as he went along. Humboldt never married and seems to have met his emotional needs through work; Bonpland was a womanizer. Humboldt was rich; Bonpland was usually broke. But they shared a love of nature, a belief in the glory of science, and a yearning for travel and adventure. The two resolved to see the world together. Their plan was to travel to North Africa, but the ship that was to carry them from Marseille to Algiers was wrecked before it could reach them, and so in the winter of 1798 they went instead to Spain, studying Iberian geography and nature en route. In Madrid their luck improved.

The Saxon ambassador to Spain was enthusiastic about science, and helped the would-be explorers win the ear of one of the Spanish ministers. This acquaintance led in turn to an audience with King Carlos IV. At the time, Spain's colonies were closed to foreigners and were permitted to trade only with Spain. No foreign scientist had visited South America for more than 60 years, and the continent's interior was essentially uncharted. Humboldt persuaded the king that Spain's dominions should be studied. With his geological expertise, he could assess the empire's mineral wealth, advise on how to get better returns from existing mines, and suggest where to look for new deposits. As biologists, Humboldt and Bonpland could collect specimens for the royal museum and gardens. Carlos IV agreed—Humboldt and Bonpland had to travel at their own expense, but were given documents that allowed them to go where they liked and instructed Spanish ships to give them passage. The king's decision was to backfire. After his South American voyage, Humboldt returned to Paris and published accounts of his journeys. Another temporary Parisian, Simon Bolivar, read of the wonders of Incan and Aztec civilizations and, between the lines, of the corruption, injustice, and neglect of Spanish colonial rule, particularly slavery. Bolivar sought out Humboldt, himself a republican and fierce abolitionist, and the two met several times. A few years later Bolivar returned home and liberated most of Latin America from Spanish rule.

The Greatest Scientific Traveler Who Ever Lived

On June 5, 1799, Humboldt and Bonpland set sail in the *Pizarro* from La Coruña at the northwest tip of Spain. As he sat in his cabin waiting to depart, Humboldt wrote a letter to a friend explaining that, rather than just cataloguing what he saw, he wanted to explain it:

> I shall collect plants and fossils and make astronomical observations. But that's not the main purpose of my expedition—I shall try to find out how the forces of nature interact upon one another, and how the geographic environment influences plant and animal life. In other words, I must find out about the unity of nature.

Humboldt and Bonpland spent the next five years exploring the Americas. They explored coastal Venezuela and journeyed 1,700 miles up the Orinoco River, mapping it as they went and showing that its upper reaches connect with the Amazon. They studied the tribes living along the river's banks, and there is an apocryphal story that Humboldt reconstructed the language of an extinct tribe from the vocabulary of a parrot he bought from a neighboring tribe. They went to the Andes, Mexico, Cuba, and the United States. The journey has been described

Alexander von Humboldt (1769–1859).
Credit: Bildarchiv Preussischer Kulturbesitz, Berlin.

as the scientific discovery of America, and it became the foundation of Humboldt's life's work. The accounts of his voyage inspired a generation of nineteenth-century naturalists. He was a hero to Darwin, who described him as "the greatest scientific traveler who ever lived" and joined the crew of the *Beagle* to follow in Humboldt's footsteps. He recommended Humboldt's writings to his friends and once wrote: "I shall never forget that my whole course of life is due to having read and re-read as a youth his *Personal Narrative* [of the American expedition]." One twentieth-century biographer wrote that Humboldt "combined meteorology, geography, geology, botany and zoology and, single-handed, created the science of ecology."

Back in Paris, Humboldt spent his inheritance—within a few years he was nearly as poor as Bonpland—on publishing his observations and ideas from the voyage. The core of his thinking about tropical diversity is contained in an 1807 essay, *Ideas for a Physiognomy of Plants*. Like Willdenow, he equated warmer climes with more species:

> The verdant carpet which a luxuriant Flora spreads over the surface of the earth is not woven equally in all parts; for while it is most rich and full where, under an ever-cloudless sky, the sun attains its greatest height, it is thin and scanty near the torpid poles, where the quickly-recurring frosts too speedily blight the opening bud or destroy the ripening fruit. ... Those who are capable of surveying nature with a comprehensive glance, and abstract their attention from local phenomena, cannot fail to observe that organic development and abundance of vitality gradually increase from the poles to the equator, in proportion to the increase of animating heat. ... It is beneath the glowing rays of the tropical sun that the noblest forms of vegetation are developed.

Tropical Wonders

Discoveries made since Humboldt marveled at the Orinoco have only reinforced our view of tropical life. There are not just more plant species in the tropics, there are more woodpeckers, ants, and monkeys—and birds, insects, and mammals in general. The same goes for parasites: Most human diseases are tropical diseases, and the tropics harbor more pathogens of all kinds. Even cultural diversity follows a similar pattern. In sub-Saharan Africa, the areas that contain the most species also contain the most linguistic groups.

Forests are not the only environment where diversity peaks in the tropics. Grasslands contain more species nearest the equator and so do deserts. More species dwell in tropical lakes and rivers than in temperate fresh water. In the sea, diversity in fish, molluscs, and plankton peaks at the equator. Tropical seas contain more species both along coasts and in the open ocean. Nor is the diversity gradient a recent thing. Fossils, including those of trees and foraminifera, the planktonic creatures beloved by D'Arcy Thompson, show that the tropics have contained more species than the temperate zones for at least the past 250 million years.

The diversity gradient is pervasive and incontrovertible. But it is general and qualitative, and there are many complications and exceptions to the general pattern. In some groups—such as aphids, seabirds, and marine mammals—diversity peaks in the temperate regions. Points at the same latitude in different continents have different numbers of species. The steepness of the diversity gradient varies between groups and between the northern and southern hemispheres. And although diversity gradients have probably been constant throughout the earth's history, the fossils suggest that they are steeper now than at any point in the past 65 million years. Nevertheless, such a widespread phenomenon cries out for a general explanation. Humboldt saw a clue as to what this might be in another pattern in diversity.

Humboldt could never see a mountain without needing to stand on its summit, and in the Andes this obsession led to an attempt on the extinct volcano Chimborazo, in Ecuador. The mountain is 20,700 feet high; at the time it was thought to be the world's highest. Bonpland and Humboldt made it up past 19,000 feet, higher than any human had gone before, until a wall of ice and snow blocked their route to the summit. When news of the explorers' feat reached Europe, it knocked Napoleon's conquests off the front pages. Napoleon was not impressed. (Back in Paris, Humboldt was presented to the emperor: "I understand you collect plants, monsieur," he said. "Yes, sire," Humboldt replied. "So does my wife," Napoleon responded.)

As he ascended Chimborazo, Humboldt decided that the bleeding gums and other unpleasant symptoms he was suffering must be due to lack of oxygen—making him the first person to diagnose the cause of altitude sickness. He also noticed that gaining altitude had the same effect on the vegetation as traveling north. As he climbed, so the tropical forest gave way to cypresses, oaks, and pines, much like the trees in European forests:

> The extraordinary height to which not only individual mountains but even whole districts rise in tropical regions, and the consequent cold of such elevations, affords the inhabitant of the tropics a singular spectacle. For besides his own palms and bananas he is surrounded by those vegetable forms which would seem to belong solely to northern latitudes. . . . Thus nature has permitted the native of the torrid zone to behold all the vegetable forms of the earth without quitting his own clime.

The key to this changing vegetation, and therefore the explanation for tropical diversity, Humboldt suggested, was water. As you move away from the tropics, or gain altitude, freezing temperatures became steadily more common. In such an environment, plant life is suspended for much of the time: "Nature undergoes a periodic stagnation in the frigid zones: for fluidity is essential to life." Humboldt believed that freezing was a stricture that relatively few forms of plant life could adapt to. Only those that could either withstand the cold or shed their leaves and wait for spring would survive. In the more permissive equatorial environment, life ran riot. "The nearer we approach the tropics, the greater the increase in the variations of structure, grace of form, and mixture of colours."

A Harsh World?

The idea that the tropical climate is more conducive to life, and so leads to a greater variety of it, is a persistent theme of explanations for tropical diversity. But it is a more slippery notion than it first appears. The idea carries a powerful whiff of circularity. The tropics have lots of species because they are a benign environment. How do we know they are a benign environment? Because they have lots of species. We usually distinguish an environment as harsh or lush by its abundance of life, but this says nothing about what in that environment might cause that abundance or scarcity.

Anyway, who's to say what's harsh? True, environments that seem

ill suited to our notion of an easy life—hot springs, salt lakes, deserts—
have few other species living in them. Yet to a thermophilic bacterium,
a hot spring is home. To a *Stenocara* beetle, which can collect the water
from fog on its wings, the Namib Desert is just fine. If these species can
adapt to their environment, why haven't more? Hot springs are rich
in nutrients, after all, and there is no shortage of solar energy in
the Namib.

There are ways to save arguments about environmental quality
from circularity. One is to replace harshness with commonness. Some
environments, such as the climates and conditions that support forests,
are widespread. Others, such as hot springs, are usually small and
isolated. We would expect lots of species to adapt to life in a common
environment, but few of them would then be equipped to colonize a
very different one. Yet some environments with relatively few species,
such as deserts and tundra, are neither small nor isolated. Their environments, however, are hostile to life by other measures than the
number of species found there. Here, it might be the physical and
chemical limits on living processes that limit diversity. Life's chemistry
needs water, so living in places where water is absent or frozen will be
difficult and expensive, requiring resources to be diverted from growth
and reproduction. But the question of why, given that a few species can
adapt to life on the tundra, more cannot still applies.

Evelyn Hutchinson addressed this problem in his *Homage to Santa
Rosalia*. Maybe, he suggested, as well as placing physical limits on what
can live where, climate controls diversity by controlling life's fuel
supply. Hot and sunny—high-energy—environments contain more
species than cold, gray places. Tropical forests also get more rain, another essential ingredient for plant life. In fact, a place's temperature
and rainfall more accurately reflect its species diversity than does its
latitude—the diversity gradient is climatic. The match between energy
and diversity stares us in the face. Perhaps this is why more species
survive in the tropics than temperate or cold climates.

In other words, the reason there is only one species of polar bear is
not that only one bear species has evolved to hunt on ice floes and
swim in the Arctic Ocean, but that there isn't enough food in the Arctic
to support two bear species. Every species must maintain a certain

number of individuals to survive. If its numbers fall too low, life's cruel lottery will finish it off. Perhaps environments with more energy can support a greater quantity of life and so allow more species to keep their numbers on the right side of the threshold for survival.

But the idea that energy promotes diversity by allowing greater numbers of organisms to live together also has its problems. Tropical seas and soils are both low in nutrients and rich in species. More resources do not necessarily mean more diversity. Adding nutrients to lakes, rivers, or soil, in the form of fertilizer or sewage, leads to a drop in the number of species. A few fast-growing species, good at sucking up abundant nutrients, come to dominate. On the other hand, this might be because few of the local species are adapted to such conditions. Over evolutionary timescales, one might expect diversity to rise as more species evolved to deal with the richer environment.

And in reality it is the quality, not the quantity, of tropical life that is outstanding. As Al Gentry found, and Brian Enquist and Karl Niklas confirmed, there aren't more trees, or a greater mass of wood, in a hectare of Amazonian rain forest than a hectare of Alaskan conifer woods—just vastly more species. Other researchers have found that the same goes for North American birds and butterflies. There is a slight trend toward increased numbers of individuals at lower latitudes, but the number of species rises far more quickly.

Diversity Through Stability

Perhaps, then, it's the quality of the climate, not the quantity of energy, that's important. As well as being warmer and wetter, the climate in the tropics is, as Humboldt noted, more stable, lacking the seasonal fluctuations in temperature found in temperate regions.

This stability might encourage diversity by letting species be more specialized, making their niches narrower and allowing a finer division of natural resources. If an animal in a temperate forest lives on fruit, or leaves, or insects, it has a problem. These food sources disappear for some of the year. The animal must either broaden its diet, hibernate, or migrate. But in the tropics it can eat leaves, fruit, or bugs all year round, leaving more resources for other species. Perhaps stability promotes diversity in general. Some other highly variable environments, such as estuaries, which alternate between saltwater and fresh water twice daily, have relatively low numbers of species.

But while the weather in the tropics might be more agreeable, the biological environment is as cutthroat as anywhere else, perhaps more so. This too has been suggested as the cause of abundant tropical diversity. Freed from the stress of coping with the inclement or unpredictable, tropical species might instead evolve to exploit each other by becoming better competitors, or predators, or parasites. If tropical species faced fiercer struggles with their enemies, or with others of their own kind, than temperate ones, they might compete less with their neighbors over resources—the force that drives species out. Thus might more tropical species coexist.

But although a stable environment or more intense biological interactions could enhance diversity, they can only be secondary factors. The problem is that nastier predators or narrower niches are *consequences* of the number of species in a place as much as they are *causes*. There is evidence that competition, predation, and parasitism are stronger in more diverse ecological communities. But this doesn't explain why there are more species there to begin with. So such arguments also carry the risk of circularity. There is evidence that diversity can beget diversity: A study by Brent Emerson and Niclas Kolm of the plants and insects of the Canary Islands and the Hawaiian archipelago found that, on islands with more species, new species are more likely to evolve. Islands with the most species also had the most endemic species, which are more likely to have evolved on that island. So the twists and turns in the struggle for life may well make the gradient in diversity steeper. But for them to have their effect, that gradient needs to be there in the first place.

A Madagascan Surprise

A recent theory, however, suggests that this gradient might appear regardless of temperature, rainfall, or anything else. In the early 1970s, Robert Colwell, then a grad student at the University of Michigan, began looking for a model of a uniform world, to show what it is about our own irregular world any theory needs to explain. Where would species be found if every part of the world had the same climate? The obvious assumption is that they would spread out evenly and that no one continent or ocean would harbor more biodiversity than anywhere else. But Colwell found that even in a world without tropics more species would live around the equator.

Colwell based his model on the geometry of species' homelands. Every species has its range: a stretch of land or water where it lives and beyond which it rarely strays, thanks to barriers such as seas, mountains, or deserts. This is why that subset of British bird-watchers known as twitchers get so excited when a hapless red-eyed vireo is blown across the Atlantic. Most species have small ranges. The giant onion, *Allium pskemense*, is restricted to the western part of the Pskem mountains in China, and bonobo chimpanzees are found only in one corner of the Congo Basin. A few species, such as barn owls, which are found on every continent except Antarctica, have spread themselves very widely. Humans, a close relative of bonobos, live where even barn owls don't go. Colwell took the distribution of range sizes—a few large and lots of small—and spread them out at random across a continent.

He saw that more species would be found in the middle of such a random continent because this is where most species ranges would overlap. And if species were spread at random over our world, more should end up at the equator, because this is where the middles of most oceans and continents are found. Colwell explains the effect by comparing the world to a pencil case. Take a pencil case and fill it with pencils of different lengths, from short stubs to long ones that barely fit in the box. The long ones represent species with big ranges, the stubby ones those with small. Shake the pencil case, so that the pencils are spread randomly along its length, but do not reach beyond the box, just as a land-dwelling species cannot live at sea. Then slice cross-sections through the box, and see how many pencils (species) are caught in each slice. The middle slices should contain more, not because the middle of the box is more pencil-friendly, but because any pencil longer than half the length of the box will inevitably be found in the middle. Species, he argues, should distribute themselves in a similar way, even without different environments.

Like neutral ecology, this model seems to be throwing biology away, by showing that the patterns in nature can arise without any of the things that scientists traditionally invoke as their causes. And like neutral ecology, Colwell's theory made many people—including Colwell himself—uncomfortable, even hostile. "At first, I only believed it on alternate Tuesdays," he says. But some spectacular matches between the model's predictions and the natural world changed his mind.

The first match was in Madagascar. Going by latitude, biodiversity in Madagascar should peak at the island's north, as this is the warmest, wettest region closest to the equator. In fact, in 1999, David Lees and his colleagues, after surveying 637 species of rain forest mammals, butterflies, birds, amphibians, and reptiles, revealed that more species are found in the middle of the island than anywhere else. Colwell was so impressed he wanted to call his theory the Perinet effect, in honor of the Madagascan reserve containing the peak of diversity, but that is a colonial name. Wanting to be politically correct and thinking the Madagascan name, Analamazoatra, too much of a mouthful, Colwell settled on the less romantic "mid-domain effect."

The mid-domain effect can also explain how diversity varies with altitude. Ever since Humboldt, most biologists had thought that the altitudinal gradient in diversity matched the latitudinal one, with a steady decline from low to high. But as the studies accumulated, they were forced to reassess this notion. About half of the counts showed, in fact, that more species are found at intermediate altitudes, neither up nor down. Climate alone cannot explain this pattern, but the mid-domain effect can because this is where there should be most overlap between the ranges of lowland and mountain species. Something similar might happen in the sea. It was long thought that diversity should decline with depth, with most species in shallow waters and fewest in the abyss. But there is some evidence that middling depths, of about 2,000 meters, are most diverse, although we know so little about marine diversity that it is impossible to draw firm conclusions.

Sometimes biology does not comply with the mid-domain effect's predictions. Mid-domain models predict that endemic species with small ranges should be spread around at random, just as the short pencils could be found anywhere in the box. Yet most small-ranged species are still found near the equator. Colwell agrees that the mid-domain effect is not the sole explanation for the latitudinal gradient in diversity, but argues that it was never meant to be. It is, he says, one more factor that must be taken into account. It is also a reminder that, before we propose an explanation for a biological trend, it helps to work out exactly what needs to be explained and gives a glimpse of what biology would look like if there were nothing biological to explain. Models such as the mid-domain effect and neutral ecology, which treat species or individuals in the same way that theories from physics treat particles, looking for the emergent properties in their behavior without worrying about the biological causes, are proving fruitful in this regard. Often the results they produce are very different from what we would have assumed—a world without ecology would not be a world without patterns.

The Importance of History

Evelyn Hutchinson described nature as an evolutionary play acted out on an ecological stage. All the ideas about tropical diversity that we have encountered so far—the school of thought that Humboldt founded—are ecological. They deal with the stage and offer reasons why the tropics might be able to hold more actors. They ignore history—what has been going on in the drama up until the point we began watching. The other main strand of ideas used to explain the latitudinal gradient in diversity looks instead to history, suggesting that the play's script has been different in different parts of the earth. To put it another way, past events might control diversity more than current conditions. Another great scientific traveler, Alfred Russel Wallace, founded this intellectual lineage.

Born in Wales in 1823, Wallace came from a family impoverished by his father's succession of business failures. Alfred had little formal schooling and left home at 13. Soon after, his elder brother William began to train him to be a surveyor. While learning his trade, Alfred became interested in geology and natural history.

When their father died in 1843, William could no longer afford to support his brother. Alfred became a schoolmaster in Leicester, a market town in the English midlands. In the town library he read Humboldt's story of his South American journey and met a fellow naturalist, Henry Bates. The two became friends and, like Humboldt and Bonpland half a century earlier, began looking for expeditions. They settled on a journey up the Amazon. Amazonian nature was unexplored, and they could finance the trip by selling the specimens they bagged to museums and collectors in London. The two left England in April 1848.

Wallace and Bates did make epic journeys in South America, but separately. No one knows why, but they parted company a few months after reaching Brazil. Wallace spent the next 4 years (Bates stayed for 11) traveling throughout Amazonia, going farther upriver than any previous naturalist, living with native tribes, and, with gun and collecting jar always at the ready, snaffling every species that crossed his path. He sent back regular shipments of beetles, butterflies, birds, alligators, monkeys, plants, and fish, to be sold by his London-based agent.

The trip, however, ended in a series of disasters. Alfred's younger brother Herbert, who had come to Brazil in 1849 to join Alfred and pursue his own career as a collector, died of yellow fever in June 1851. Alfred suffered recurrent bouts of fever, some so bad he feared they would kill him (he would carry malaria for the rest of his life). And three weeks into the voyage home, after leaving Brazil in July 1852, his ship, the *Helen*, caught fire and sank. Wallace and the crew spent 10 days in a small boat, eking out what little food and water they had, before a passing ship came to the rescue. Alfred had lost his journals, his drawings, a small menagerie of live animals, a commercial collection that he estimated to be worth £500, and his personal collection, containing "hundreds of new and beautiful species, which would have rendered my cabinet . . . one of the finest in Europe." All that survived was one parrot that had made it off the burning ship.

Wallace had hoped to return to London with a collection that would secure his financial future and make his scientific name. As it was, he could only be grateful that his agent had insured the lost specimens. There was nothing to be done except set off for the tropics again. Eighteen months later, in January 1854, he went east. Wallace spent the next eight years wandering throughout the Malay Archipelago, the region that is now Indonesia and Malaysia, from Singapore to New Guinea. Again, he aimed to fund the trip through commercial collecting. It was the basis of all his future biology and, he said, "the central and controling incident" of his life.

Wallace did not give a full account of his ideas about diversity gradients until much later, in an 1878 book called *Tropical Nature*. In it he gives a beautiful description of what it feels like to be a temperate naturalist in a tropical forest:

> Instead of endless repetitions of the same forms of trunk such as are to be seen in our pine, oak or beechwoods, the eye wanders from one tree to another and rarely detects two columns of the same species. . . . If the traveler notices a particular species and wishes to find more like it, he may often turn his eyes in vain in every direction. Trees of varied forms, dimensions and colors are around him, but he rarely sees any one of them repeated. Time after time he goes towards a tree which looks like the one he seeks, but a closer examination proves it to be distinct. He may at length, perhaps, meet with a second specimen half a mile off, or may fail altogether, till on another occasion he stumbles on one by accident.

Wallace subscribed to Humboldt's view that climate limited the spread of plant species and that the tropics harbored those too delicate to survive harsher climes. But a great deal else had happened since Humboldt had seen the rain forest. Geologists had realized that the world was vastly older than the biblical chronology could account for. And Darwin and Wallace had separately realized that all organisms were locked in a perpetual struggle that, through the mechanism of natural selection, could create and modify species. Even in Leicester, Wallace had been thinking about how species might give rise to new species. He continued thinking about the topic in the Amazon, discussing the subject with Bates and other Western naturalists he encountered on the river. Full revelation came in early 1858.

Collecting on the island of Gilolo (now Halmahera), which lies between Sulawesi and New Guinea, Wallace fell sick with malaria. Lying in bed with a fever, his mind turned to Thomas Malthus's theories of how war, famine, disease, and disasters kept human numbers in check. Something similar, he thought, must apply to other animals. He recalled his next thought in his autobiography, published in 1905:

> Then it suddenly flashed upon me that this self-acting process would
> necessarily *improve the race*, because in every generation the inferior would
> inevitably be killed off and the superior would remain—that is, *the fittest
> would survive.*

When his fever eased, Wallace wrote up his ideas in a short paper,
"On the Tendency of Varieties to Depart Indefinitely from the Original
Type," and sent it to Charles Darwin for comment. Unbeknown to
Wallace, Darwin had had a similar insight 20 years previously, also
inspired by Malthus, but had never published (one of science's great
what-ifs is how biology would have developed had Wallace perished in
the mid-Atlantic with the *Helen*). Darwin, poleaxed by Wallace's mis-
sive, hurried to secure his claim to the idea. On July 1, 1858, at a meeting of the Linnean Society in London, (he was not consulted) Wallace's
paper and a previously unpublished 1844 essay by Darwin were
presented jointly, and natural selection was unveiled.

Guided by his evolutionary thinking, in *Tropical Nature* Wallace
suggested several reasons for the diversity gradient. In a uniform
climate, he reasoned, competition between species would be stronger,
resulting "in a nice balance of organic forces, which gives the advantage now to one, now to another species, and prevents any one type of
vegetation from monopolising territory." And species could also be
more specialized, to shade or sun, or to grow on other plants as
epiphytes or parasites: "Every place in nature [is filled with] some
specially adapted form."

As well as being the codiscoverer of natural selection, Wallace
founded what is now called biogeography. This discipline combines
elements of ecology, paleontology, evolutionary biology, and geology
to investigate how species arise, disappear, come together, and move
apart under the influence of processes such as continental drift, volca-
nic eruptions, and climate change. Famously, Wallace noticed that the
Indonesian islands of Lombok and Bali, although separated by only 25
kilometers of sea, have very different flora and fauna. This divide became known as the Wallace Line, and it marks the boundary between
the Asian and Australasian tectonic plates. The islands are only recent
neighbors. Most of their species arrived when the two were far apart,
and there has been little interchange since they drifted together.

In a similarly biogeographical vein, Wallace also argued that geological history could account for the greater diversity of tropical animals. Successive ice ages have smothered temperate species under glaciers, "destroying most of the larger and more specialised forms . . . [and] necessitating the commencement of the work of development in certain lines over and over again." In the tropics, "evolution has had a fair chance" and, unchecked by periodic catastrophes, was able to accumulate the full, thrilling range of forms, sizes, colors, and behaviors. The tropics are more diverse because they are older, and so provide both a laboratory for producing evolutionary novelties and a preserve where the results of these experiments can survive.

Where disaster strikes, we do see large, and long-lasting, reductions in species' diversity. In 1883 the eruption of Krakatoa sterilized several islands in its vicinity. These islands are green once more, but it takes time to recover from such a catastrophe, and they still have fewer species than similar islands not caught in Krakatoa's blast. And there is evidence that the most recent ice age, which ended about 12,000 years ago, still casts its shadow over the spread of species. In North America the areas most recently covered by ice harbor fewer bird species, and the comparative youth of the Great Lakes of America may be one reason why they contain vastly fewer fish species than African lakes such as Victoria and Malawi. Pollen analysis shows that tree species took millennia to recolonize American forests in the wake of the glaciers' retreat. History has been invoked to explain other differences in species diversity besides that between the tropics and the temperate zone. North America has more tree species than Europe because, it has been suggested, Europe's mountains—the Alps—run east to west. As European trees retreated south before the advancing glaciers, they would have been squeezed between the ice and the mountains, and many would have gone extinct with their backs up against the Alps. American trees faced no such barrier, as the Rockies run north–south, and so could have gone and returned more easily.

Past history, then, is an important factor in current biodiversity. Time is needed both for existing species to colonize places depopulated by a catastrophe and for new species to evolve to exploit new habitats. The longer since a place was wiped bare, the more species one would expect to see there. But although this reasoning helps explain local differences in species diversity, it is less clear whether history—like all the other explanations encountered so far—can explain the global equator-to-pole gradient.

Glaciers do not inevitably wipe out all species in their path. Rather than being eliminated, the temperate zones and their species might simply have shifted to follow the climate. This should be particularly true in the sea, where life is more mobile. And the climatic changes wrought by ice ages also affect tropical forests. Fossil pollen shows that the extent of tropical forests has varied a great deal. It appears that drier and warmer climates lead to large patches of grassland appearing amid the trees. It's also been argued that isolating clumps of forest from one another promoted diversity, because new species would be more likely to evolve in forest fragments, just as they do on islands. But you can't have it both ways—either stability promotes greater diversity or it doesn't. Current conditions seem a better guide to current diversity than past history—the number of species found in a place is correlated more closely to its present climate than to the length of time since that place was last covered in glaciers. There seems to be some balance that, given a few thousand years, restores itself. In the Cretaceous period the world's climate was much milder. Yet the diversity gradient was still in place. And mass extinctions have hit the tropics just as hard as the rest of the world. It is hard to argue that any one environment is older than any other.

Big Places Have More Species

But if the tropics are not biodiversity's museum, they might be its cradle. There is evidence that, even if they are just as likely to die out near the equator, species are more likely to be born there. Two reasons have been suggested for this phenomenon: the tropics are larger, and they are warmer.

First, larger. It doesn't always look it on maps, because the Mercator projection typically used to transform the spherical Earth into a two-dimensional map squashes land at the equator and stretches it at the poles, making Greenland look nearly as big as Africa. But looking at a globe, it is clear that the tropics are by far the largest climatic zone on

Earth. They cover about 55 million square kilometers, three times more than the next largest zone, the tundra. Likewise in the sea, tropical oceans cover more of the earth's area than any other land or water mass.

What's more, all the world's tropical habitat is in one continuous swath lying on either side of the equator. Other zones, such as temperate or subtropical, are found in separate bands in the north and south. A grass seed or a ground squirrel trying to migrate from the North American prairie to the South American pampas has got to cross thousands of miles of hostile tropical habitat. But a monkey could, in theory, travel from Mexico to Argentina without entering alien territory. And within the tropics, between about 25° latitude north and south, the climate is relatively constant. The annual average temperature is about 28°C throughout this zone. As you leave the tropics, the temperature drops by about 8°C for each 10°N you travel. So species adapted to tropical climates have much more land or water to play with compared with their nontropical counterparts.

It's obvious that larger areas will contain more species. They can harbor more individuals and a greater diversity of habitats. The relationship between a place's area and the number of species found there follows a power law. The exponent in the species-area power law—the bit that is 3/4 in Kleiber's rule—varies from place to place but is usually between 0.1 and 0.25. So the rate at which you find new species slows down as the area you search expands. Measuring the form of this power law also allows conservation biologists to make predictions about the effects of habitat destruction. For example, the species-area curve for the forests of Eastern America suggests that removing half of the forest will drive about one in every seven bird species to extinction. Sure enough, half of the forest has been lost since 1870, and in that time four of the 30 bird species found there and nowhere else have disappeared.

Less obvious, from this line of reasoning, is why the tropics should also contain, as they do, *proportionately* more species than the temperate regions, a greater number for each square kilometer of area. John Terborgh and Michael Rosenzweig argue that the tropics' large area creates a high diversity because tropical species, given more land, have larger ranges—and that this makes them both less likely to go extinct and more likely to give rise to new species. Tropical species are less likely to go extinct because a species spread over a large area is unlikely to succumb to any single fire, flood, hurricane, or epidemic. Widely spread species are also more likely to bud off new species, because small pockets of their populations will become isolated. Such pockets will stop interbreeding with the rest of the species and so become genetically different from their neighbors. If the isolation lasts long enough, they will no longer be able to breed with the parent population, and a new species will be born. A large range is more likely to be fragmented by inhospitable habitat, be it rivers, mountains, grassland, forest, desert, or sea, creating the isolated populations that give birth to new species.

Rosenzweig argues forcefully that any study of what controls biodiversity must take account of area. He has collected evidence to support the idea, comparing environments with similar climates but different areas, and showing that the larger provinces contain proportionately more species. The Amazon River basin, for example, is about 1.5 times the area of the Congo basin but contains well over twice as many fish species, 1,300, compared to the Congo's 500.

But other ecologists have raised objections, and pointed out exceptions, that make it unlikely that the tropics' great area is a general explanation for their great diversity. Not all habitats are more widespread in the tropics. There is just as much fresh water elsewhere, yet most freshwater fish are still tropical species. Coastal shelves, the shallow seas that border the continents, have no greater extent in the tropics than elsewhere. Yet they harbor a far greater diversity of molluscs, to name one group investigated by biologists.

It is also not clear whether tropical species do have especially large ranges. Some ecologists have argued that species living in seasonal temperate climates should roam farther, because the adaptations that let animals and plants cope with wintry weather should also equip them to live in a wide range of different places. And how far a species can spread will depend in part on how many other species it must compete with, so, in addition to range size being one thing that might influence diversity, diversity will also influence range size. It is fearsomely difficult to disentangle the causes and consequences of diversity. Nor is it certain that species with big ranges will give birth to more new species. A species with a really large range will be less likely to become divided.

Instead of cutting a population in two, a mountain range or a forest could just become an island of inhospitable habitat in the sea of that species' range.

Evolution's Workshop

That's how being larger might make the tropics more diverse. Now for warmer. As we have seen, the amount of energy pouring into a place is a good guide to the number of species living there. But it seems that the amount of fuel alone cannot account for the difference in diversity. Instead, extra energy might create species by speeding up life's tempo. This is where metabolic rate re-enters our story.

By influencing metabolism, temperature affects evolution. Organisms whose body temperature matches that of their environments—meaning plants and all animals except for the warm-blooded birds and mammals—have faster metabolic rates in warmer climates and so grow more quickly. They will reach maturity quicker, which will make each generation shorter, and so natural selection will have more to work with. The malaria-carrying *Anopheles* mosquito, which can crank through 10 generations in a year, is likely to evolve more quickly than a temperate species that can only manage one. In general, species with fast generation times can adapt quickly. That is why pesticide resistance in insects, and antibiotic resistance in bacteria, is such a problem.

Jim Brown's team has found that the warmer and smaller a species is, the more quickly its DNA changes. Like cellular metabolic rate, mutation rate over time falls as the $-1/4$ power of body mass. And besides speeding up the life cycle, hot temperatures cause mutations. Male mammals carry their testicles outside their bodies to keep them cool, in an attempt to reduce the mutation rate in their sperm. The effects of temperature on the pace of life, via metabolism, therefore provide a mechanism by which climate can influence the rate at which new species form. About the same amount of energy is needed to spark a mutation in all cells and species. The effect of temperature on mutation does not translate directly into evolutionary change. How much a species alters through time depends on the selective forces it experiences. But it does show that, other things being equal, the DNA of small, hot, fast-burning plants and animals will change more quickly than that of large and cool living things. Metabolism, via its affect on the rate of life, provides a mechanism by which climate can influence the rate at which new species form.

Brown's team used temperature and metabolic rate to build a model of diversity variation with temperature, based on temperature's affect on metabolism (note that this says nothing directly about the rate of mutation or the appearance of new species). Patterns in the diversity of trees, amphibians, freshwater fish, marine molluscs, and the parasites of marine fish all seem to match the model's predictions, with a place's diversity rising as it gets warmer.

Fossil evidence supports the idea that the tropics are evolution's workshop. David Jablonski has shown that most groups of marine animals first appear in equatorial rocks. Many temperate species—including us—are tropical migrants, not creatures born and bred in the cold. Tropical rocks also contain more young species than temperate rocks, suggesting that evolution works faster close to the equator. And from looking at fossil foraminifera, Drew Allen, a former student of Brown, has found that tropical species, which have metabolic rates 15 to 20 times those of their polar cousins, evolve new species more rapidly than their cold-water counterparts. Tropical forams also go extinct more quickly, perhaps because if one species splits into two, each new species has a smaller population. By working out at the molecular level how much energy is needed to cause a mutation and the amount of time it takes for a new species to evolve, Allen has made a back-of-the-envelope calculation that it takes 10^{22} joules to evolve a new species of foraminifera. About the same amount of energy shines down as sunlight on Earth each day.

The link between temperature, metabolism, and evolution is a good candidate for an explanation of why there are more species in the tropics. It is one of the few relationships between environment and biology that ought to hold for every species, wherever it lives; many of the other explanations were derived from observing land plants and animals, and they founder when applied to marine life. It seems to fit with fossil evidence. It predicts roughly how many species we should expect to find in a place of a given temperature—rather than just offering a reason for the shape of the trend—which allows its ideas to be tested. Our understanding of how temperature, via metabolic rate, influences development and mutation provides a mechanism for more rapid tropical evolution. And in some senses it looks good by default. Most other reasons for the diversity gradient have been around for a while, giving people time to point out limits to their generality. Ideas about evolutionary rates are relatively new and have yet to be ground down by the mill of academic scrutiny.

No Easy Answers

After two centuries of research, a toolkit of ideas (or a bunch of hedged bets) is starting to emerge that might enable us to explain why there are more species in one place than another and why there are more species in the tropics than anywhere else. Climate supplies energy, which influences evolution. Climate also influences the harshness and productivity of different environments, which controls what species can live there. History, in the time needed for diversity to recover from past shocks, is surely a factor, as is the world's geometry, because the tropics are larger than anywhere else and because they are in the middle of their domain.

How climate, history, and geometry might interact to control biodiversity is still poorly understood and controversial. Their combined effect is bound to be complex and will vary from place to place. It may be that the simple, top-down approach favored by macroecologists can only get us so far. One possible way forward is through computer simulations. Satellites provide the raw material for such models, in the shape of data on temperature, rainfall, solar radiation, and vegetation cover, which can then be matched to patterns in diversity, particularly the diversity of well-studied groups such as birds and mammals. These models will not be simple, like the 3/4 power law. They will be more like the models that climate scientists use to predict the movements of air and water around the atmosphere. These simulations require vast computer power and give ambiguous results, with sizable errors. Some ecologists are already employing such models to see how their theoretical predictions match up with reality and to forecast how species might respond to climate change. Perhaps the Alexander von Humboldt of the twenty-first century will set sail on a computer.

10   A NEWTON OF
THE GRASS BLADE?

I N THE FIRST SENTENCE OF *On Growth and Form*, D'Arcy
Thompson enlists the eighteenth-century German philosopher
Immanuel Kant in his cause, quoting Kant's assertion that without
mathematics a discipline may be a science, but it cannot be Science.

Kant certainly thought that natural philosophers should seek
unities. He complained that Linnaeus's classification system for species
focused too much on the details and ignored the whole from which
life sprang. His thinking was a major influence on Alexander von
Humboldt. But Kant, writing in his *Critique of Judgement*, doubted
that living things would give up their secrets:

> [I]n terms of merely mechanical principles of nature we cannot even
> adequately become familiar with, much less explain, organized beings and
> how they are internally possible. So certain is this that we may boldly state
> that it is absurd for human beings even to attempt it, or to hope that
> perhaps some day another Newton might arise who would explain to us, in
> terms of natural laws unordered by any intention, how even a mere blade
> of grass is produced.

Kant believed that biology and physics occupied different intellectual worlds. This book has been the story of some of the scientists who

have tried to prove him wrong: Humboldt, Thompson, Max Rubner, Alfred Lotka, Ludwig von Bertalanffy, Evelyn Hutchinson, and Robert MacArthur, the modern generation of macroecologists and their physicist colleagues. Many have invoked Newtonian science as their model and goal, and many have sought to go beyond a single blade of grass to find and explain patterns in the living world that encompass the whole planet.

No one now disputes that plants, animals, rocks, and galaxies are made from the same stuff and are subject to the same physical laws. This fact has become trivial. What is not trivial is whether we can find general laws, similar to those that physics applies to particles and planets, which explain how the living world works.

Of course, biology already has one all-pervading principle. When Kant invoked Newton, he was arguing that the world only made sense "as the product of an intelligent cause." Biology has already shown, through the theory of evolution, that this is not so. From the facts of genetic mutation, inheritance, and natural selection, evolutionary biology shows us how and why life came to be the way it is.

Many biologists believe that, other than evolution, generality in biology is unlikely. They argue that biological systems are more complex than physical ones. Particles such as electrons and quarks, for example, can be described by their mass, velocity, and electrical charge. Geologists studying the earth's crust must come to terms with a few thousand types of mineral. But there are millions of species, each differing in size, shape, behavior, habitat, range, and rarity. Within species, every individual is different, and all change over the course of their lives. Hindsight, these scientists argue, might show us how nature as we see it came about, through past events such as drifting continents and changing climates. And on a case-by-case basis, most biologists are confident that we can explain, by looking at factors such as the availability of food, the threat of predators, the fights between males, or what mates females prefer, why a particular species does what it does. But many also think that the interactions between living things and their environment and each other is too complicated, and too dependent on the circumstances of each moment, to allow generalization.

From this viewpoint, trying to find physics-like laws of nature is like trying to track down the butterfly whose wing beat caused the typhoon.

Most of the science I have written about sidelines history. Ideas such as D'Arcy Thompson's "diagram of forces," niches, network models, and energy equivalence assume that there has been enough time, and evolutionary flexibility, for evolution to find a solution to a problem—for nature to find a balance—that can be explained without reference to an organism's past or the constraints on it. The successes of such an approach show that sometimes this assumption is justified. Robert MacArthur explicitly sought nonhistorical explanations: "The ecologist and the physicist tend to be machinery oriented," he wrote, "whereas the paleontologist and most biogeographers tend to be history oriented."

But a historical viewpoint is as critical in understanding life today as it ever was. As an example, imagine the following thought experiment about how some animal groups become more diverse than others. You release a small animal species and a large one—a rat and a deer, say—onto an island with no other animals, and monitor the evolutionary results for a few million years, until many more species have evolved. Which of the original species will give rise to the most descendents? You might expect that the small animal would split into more new species because, as we have seen, there are more small niches and small animals are more diverse. The small animal would evolve more rapidly. Yet DNA evidence of some animals' evolutionary history gives another picture. Andy Purvis and his colleagues have found that small, diverse groups such as rodents do not seem to have split into new species, and filled niches, any more quickly than large mammals such as primates have. But small mammals did emerge relatively unscathed from whatever killed the dinosaurs off 65 million years ago, and so the reason there might be more of them today is that they made up a greater proportion of the ancestral stock from which today's mammals evolved. They won the race not by being faster but by having a head start. However much we might be able to explain about ecology using concepts that ignore an organism's history, the historical perspective is still crucial to understand how today's nature came to be.

Physics Envy?

Biology's would-be Newtons often get accused of physics envy. Why, say their critics, should biology be like physics? What reason is there to apply Newtonian goals and techniques to biology? Just because physics is an older science than biology does not mean that it should be a model for how to do science. Francis Crick wrote that "the ultimate aim . . . is to explain all biology in terms of physics and chemistry." But others, such as Ernst Mayr, have argued that it is not being vitalist to believe that biology has concepts that physics cannot explain. There is no physical equivalent, for example, of inheritance, behavior, sexual reproduction, or immunity. Few would argue that these were not valid scientific concepts. Instead of feeling inferior, or trying to annex biology as a province of physics, perhaps we should just accept the difference and see the two fields as equally powerful and legitimate.

Lawrence Slobodkin was a graduate student of Evelyn Hutchinson. He alerted Robert MacArthur to the potential for applying mathematics to ecology. But since the 1960s, Slobodkin has come to believe that the search for physics-like theories in biology is misguided. Biologists try to find them, he says, because their emotional attachment to nature makes them want to aggrandize their science. "Physicists are not in love with atoms, but ecologists are in love with organisms," he says. "Devising theories gives biologists an excuse for the love affair." Other biologists think that searching for generality diminishes their science more than it aggrandizes it—that it is arrogant to try and force the diversity of life, the story and traits of every species, into a simple uniform framework, and that if you do the result is drab biology. But then a hundred years ago, people used to think that the same criticisms applied to using mathematics to describe sea shells.

Some physicists who have made the journey to ecology argue that the differences between physics and biology are not as great as they appear. Robert May trained as a theoretical physicist and then made the switch to ecology, during what he calls the "romantic phase" dominated by Robert MacArthur's ideas. May's mathematical tools were ideally suited to the ecology of the time, and he went on to become an eminent theoretical ecologist, discovering, among other things, that the size of a population could fluctuate chaotically even without any environmental changes. Nature could look random even if we know it is not, and it can be unpredictable even if we know the rules. (He has since served as the British government's chief science adviser and as president of the Royal Society and is now Lord May of Oxford.) To believe, he says, that physics offers the one pure, rational method of doing science, building a theory by moving from hypothesis to experiment to refutation, is a fiction. There are whole fields, such as astrophysics, without experiments and whole fields, such as string theory, without a single data point. "Ecologists who hanker for the precision of physics don't have the faintest clue what physics is really like," he says.

Others, on the other hand, believe that it is not biology and physics that are different but biologists and physicists. The physicist Freeman Dyson wrote that scientists are split into unifiers and diversifiers and that biologists tend to be the diversifiers, "happy if they leave the world a little more complicated than they found it," and that biology lacks generality because biologists do not look for it.

What Makes a Theory

This is getting out of hand. Trying to work out the differences between biology and physics, and biologists and physicists—and arguing whether biology is in its Copernican, Keplerian, Newtonian, Einsteinian, or whoever-ian phase—is a fun coffee-break conversation for academics. But more productive than asking whether biology is, or should be, like physics is to ask whether there are simple universal laws that apply to biological systems, and whether these laws can be rooted in principles and approaches borrowed from physicists, either directly (such as the theory that network structure controls metabolic rate) or by analogy (the idea that large groups of plants and animals behave like large groups of particles). And if there are, what should we expect from them?

For starters, any law of nature should describe a pattern that would allow us to see generalities and shape our expectations. Kleiber's rule predicts that if we encounter an unfamiliar mammal, we should expect its metabolic rate to be roughly 70 times its body mass raised to the power of 3/4. This is equivalent to those laws of physics that are formal statements, such as "force equals mass times acceleration." Biologists have spotted plenty of similar patterns, such as the latitudinal gradient in species diversity, and Bergmann's rule; all have exceptions. But biological patterns such as Kleiber's rule are descriptions—the pattern the rule describes is just a more mathematical version of the argument that, because every raven I have seen so far is black, all ravens are black. Experience has led me to believe that blackness is a general property of ravens—just as it has that mass-to-the-power-of-three-quartersness is a property of metabolic rate—but I have not worked out a reason why ravens have to be black. Formulating such a rule from observation and experience begs the question of *why* living things should be like this.

Spotting a pattern therefore usually leads to a search for an explanation, perhaps in terms of evolutionary advantage, or the physical and historical constraints on evolution, or in what is statistically probable. With regard to metabolic rate, West, Brown, and Enquist's fractal network theory looks at both the best solution to a particular problem and the physical limits on that solution. It argues that metabolic rate is proportional to mass to the power of 3/4 because this is the quickest and cheapest way for a body to supply its cells with energy, thanks to the way that the geometry of transport networks changes with body size.

The network model also provides an abstraction. Theories are tools for thinking. As well as offering descriptions of the world, they seek to reveal something deeper that may not be obvious to observers and measurers. Physicists make theories by pretending that things are simpler than they really are. They understand their laws by thinking about imaginary systems, such as a universe containing one planet that orbits one star, or motion without friction, or collisions that rebound with perfect elasticity. Some ecologists argue that their science already has principles such as this. Lev Ginzburg, for example, argues that Malthusian population growth is similar to Newton's first law of motion in that it describes what will happen if nothing else is happening. A population in an unlimited world will grow exponentially; a body given a push will carry on in that direction until some other force stops it.

In biology such abstraction can have surprising consequences. Neutral ecology and the mid-domain effect show how far we can get by ignoring the things that have traditionally been the stuff of biology and hint that we may not need a full picture of the complexity of living things to understand the patterns in nature. Even in a featureless world, randomly filled with identical species, not all would be equally common or equally widespread.

This is perhaps the part of physics-type thinking that comes hardest to biologists. And with good reason. Unlike planetary orbits, living things are responsive, flexible, and adaptive. They bend the rules. Combined with the importance of history and contingency, this property limits the predictive power of biological theories and creates the long-disputed territory between life's generalities and its detail— and between the scientific lumpers and the splitters. The debate goes something like this:

> *Macroecologist:* Look, insectivorous mammals, such as shrews, have pointier faces than herbivores, such as mice. This seems like a general rule about nature.
>
> *Skeptic:* But what about the long-nosed potoroo, a Tasmanian marsupial that looks a bit like a cross between a large rat and a tiny kangaroo? It has a pointy face, but feeds mainly on fungus.
>
> *Macroecologist:* Nevertheless, my rule is good, even though it doesn't work for the long-nosed potoroo—on average, if you use a large data set, most of the herbivores are flat-faced.
>
> *Skeptic:* Phooey, your rule is no good *because* it doesn't work for the long-nosed potoroo. What's so general about a principle that doesn't apply to anything in the world as one finds it?
>
> *Macroecologist:* Well, perhaps my general principle can help you discover why, despite its fungal diet, the long-nosed potoroo has such a pointy nose.

And so on. Much of the disputed territory is a question of taste: whether it is the diversity or unity in nature that we find striking, whether we think complicated problems are likely to have complicated answers, whether we enjoy describing details or dreaming up abstractions. I have concentrated on the search for generality in nature, but there is much to be said for the alternative view. As Robert May says: "The way that evolutionary forces interact with environments is so varied that it may be a mistake to look for general explanations. The differences might be more interesting than a grand unified theory that isn't particularly grand or unified." Interesting is the key word—ideas persist because scientists find them stimulating and useful, rather than because they are grand.

One consequence of biological theory's limited powers of prediction is that we cannot expect the theories to provide catch-all prescriptions for conservation. Throughout this book I have tried to illustrate how ideas about how nature works might help us conserve nature. It has emerged that large-scale generalities such as metabolic scaling and energy equivalence are only good as rules of thumb, first approximations that can give conservationists an idea of where to start, suggesting what species might be most at risk and what conservation efforts might be most effective. "Large species are more at risk of going extinct" is useful but not decisive. A doctor confronted with a coughing patient would not make a diagnosis without a more thorough investigation.

The analogy between medicine and conservation shows us the limits of prediction. Our knowledge of the human body, how it responds to its environment, and what goes wrong with it is far more detailed than our equivalent knowledge of nature. Biomedical research is much better staffed and funded than conservation. Yet none of us can know what our health will be like in five years' time. Likewise, there is as yet no way of making precise local predictions about the effects of climate change or invasive species. On a larger scale, however, both medicine and nature are more predictable. Everyone knows someone who knows someone whose Uncle Bert smoked 60 cigarettes a day but lived to be 95. But, on average, smoking is a reliable way to reduce life expectancy. Similarly, using models of climate change's affect on species' habitats, ecologist Chris Thomas and his colleagues have predicted that, in a worst-case scenario, as many as a third of species with small ranges will be "committed to extinction," (i.e., doomed) by 2050. But what species they will be, and the knock-on effects of removing species, are unpredictable.

Finding a unity of nature would not make studying the details of nature obsolete. Indeed, finding unity depends on understanding the details. The variability of life means that in biology the ability to generalize is not enough. If you've measured one electron, you've measured them all, but, as I saw in Costa Rica, to understand a forest you must be able to see the trees, and that takes a botanist. Thinkers such as Humboldt, Darwin, and Wallace gained their understanding of how nature works from years of intimate experience of nature in the flesh and the leaf. And yet they were not just interested in what their senses told them; they also tried to abstract and unify. This combination of attributes—intrepid and reflective, naturalist and mathematician— strikes me as rather rare, and becoming more so. These days scientific lone wolves such as D'Arcy Thompson are almost extinct, and it would take a truly awesome polymath to acquire the necessary suite of skills in natural history, ecology, mathematics, and physics to devise a theory as complex as fractal networks. It may be that, in today's scientific landscape, groups of complementary specialists are better placed to find theories of nature.

One thing that helps scientists to abstract and generalize is the ability to think on scales alien to human experience, to imagine the universe in terms of immense—or infinitesimal—distances, times, and numbers. Large scales, of area or differences in body size, are also those on which the patterns in nature seem most striking. Astronomers and particle physicists deal with this scale all the time, but again it is alien to biologists. Nature is all around us, and it is easy and fun to study. I could go back to Oxleas Wood tomorrow morning, watch some robins and come up with an idea about why they nest and feed where they do. This approach has gotten biology and ecology a long way. But it may not be the best way to find generalities. As Jim Brown observed, trying to work out how a large, complex system of interacting parts like a woodland works by looking at its constituent parts in isolation is thankless. It's like trying to learn watchmaking by looking at a pile of dismantled cogs and springs. You might get an idea what each individual bit does, and some of the parts would fit together in suggestive ways. But it would be a long slog before you could build anything that kept time. Physicists know that systems with an intermediate number of parts are the most analytically intractable. In ones and twos, fine; in huge groups, statistical regularities emerge, although the behavior of each individual particle is unpredictable. Ecologists, or at least those who are trying to understand why species live where they do and why some are common and some rare, have typically focused on the intermediate scale, which is the most difficult to understand. Biology resists physics-type theories partly because at human scales there are few physics-type patterns to theorize about.

One of the reasons that general ecological ideas are in vogue is that experiencing nature on large scales now no longer needs great leaps of the imagination or self-sacrifice. To study the moon, or the rings of Saturn, one need only look to the heavens. But to truly appreciate what nature is capable of, you have to go to the tropics, which for biologists from temperate climates for most of history has meant spending months in a canoe. Now, you can jet into San Jose, rent a car, and see half a dozen different types of forest in a few days. And besides Costa Rica, the Enquist group is running projects in Mexico and Colorado. Simply staring out of a plane window is liable to set you thinking about nature on the grandest scales. Things look simpler from a distance and complex when they are in our faces.

Similarly, biologists have a newfound ability to analyze nature on a large scale. Thanks to the efforts of researchers such as Al Gentry and Steve Hubbell, scientists can call up the vital statistics of millions of trees from around the world at the touch of a button. Theorists have bigger, better, and more accessible data sets to work with, more computer power, and more sophisticated statistical methods. They no longer need to spend years testing a model. Just as genomics researchers can mine sequence databases for ideas about how genes work, ecologists can mine data about forests or food webs and get an instant idea of how theory and fact match up. It is much easier, to paraphrase Alexander von Humboldt, to survey nature with a comprehensive glance and abstract your attention from local phenomena; or as Brian Enquist once said to me: "If I didn't want to, I need never go into the field again."

Mutual Dependency

Theory and fieldwork need one other. It's obvious that theories should be tested by data. But facts also need theories, to provide a context for data, allowing facts to be linked and placed. An allometry line joins the dots between data points, turning a mass of unruly facts into a tidy trend and a simple equation. Theories also filter facts, showing how much attention and credibility each piece of data deserves and whether you should perhaps mistrust your experiment or your theory. Unfortunately, wrong theories are almost as effective at filtering facts as right ones, and once a theory is established, it takes a lot of contradictory facts to bring it down. Max Rubner's surface law relating metabolic rate to body size was dogma for half a century, and it took the combined efforts of Max Kleiber, Samuel Brody, and Francis Benedict to make it controversial again.

But once the surface law had fallen out of fashion, nothing came along in its place. As far back as the eighteenth century Antoine Lavoisier, the first to measure metabolism, realized the dangers of a theoretical vacuum. He was writing about physics, but his words should be a warning to any scientist who thinks that a tricky problem will be solved by just one more experiment: "While the spirit of system is dangerous in the physical sciences, it is equally to be feared that the disorderly heaping up of a great many experiments will obscure science rather than clarify it, raise barriers to those who wish to advance beyond the first stages, and cause long and difficult research to lead to nothing but disorder and confusion." Many of the biologists who studied metabolic rate ignored Lavoisier's warning. More and more measurements accumulated that, without a theory to beat them into shape, became a formless, incomprehensible mass.

Another thing that theories ought to do is make predictions, even if they are quite imprecise, either of what is impossible or of what will happen. Then, if the apparently impossible happens or the inevitable does not, the theory can be changed or discarded. Such predictions should apply to problems different from that for which the theory was originally devised. Yet contradictory evidence is rarely enough to do away with a theory, and biological hypotheses, such as those about metabolic rate or patterns in species diversity, are much harder to get rid of than they are to create. A biologist might create a theory to explain what he or she has seen on the prairie, or island, or rain forest, or in cows or trees. Inevitably, it will fit that case. But its failure to explain a similar pattern elsewhere is rarely enough to kill it off. Brian

Enquist has not given up on metabolic ecology because it does not explain why forest respiration, as revealed by FLUXNET, does not vary with temperature. Instead, he and Drew Kerkhoff are looking for what else might be going on that can reconcile their theory with the data, in the same way that Chris Carbone and John Gittleman worked out how the rules of energy and population density apply to carnivores. But without some pruning, ideas will pile up, each good at explaining some things and weak at others.

Life may actually be like this, with an atlas of ideas needed to describe the different mechanisms operating at different scales and in different places. But biologists will only find generality if they look for it. Could a biologist make a prediction like the one based on Newtonian theory, when wobbles in the orbit of Uranus were used to predict the existence and position of Neptune, or like Einstein's prediction that large masses would bend starlight, later proven true during observations of a solar eclipse? What would such a biological prediction be like? Conservation efforts will help here: When trying to apply an idea in aid of a definite goal, it really matters whether that idea is right or not.

On my suggested criteria—description, explanation, simplification, and prediction—the fractal network theory of metabolic rate does rather well. It takes a pattern in nature and argues that the pattern is a consequence of how the laws of physics limit the workings of living things. In other words, it uses mechanical principles to explain the internal possibilities of organized beings. It provides an abstraction: This is how animals and plants should work in the ideal case and in general, not in every specific instance. The theory then goes on to make many more predictions about other aspects of biology, such as anatomy, growth, development, reproduction, population density, species diversity, and rates of evolutionary change. You would hope that if these relationships were drastically different from the quarter-power laws that the fractal theory predicts, it would keel over, but none so far have been.

The fractal network theory also exploits new mathematical tools that seem tailor-made for biology. Another reason that few biologists have risen to Kant's challenge is that, even if they started thinking like physicists, classical Newtonian physics and mathematics are unequal to the task of describing living things. As Lawrence Slobodkin says: "Almost any prepackaged mathematical structure used in biology was designed for something else. Biology wears these clothes like hand-me-downs."

The fields of chaos, complexity, and emergence being studied by researchers at the Santa Fe Institute and elsewhere came about through the efforts of biologists such as Ludwig von Bertalanffy to think like physicists, and of the biology envy of physicists such as Robert May and Geoff West. As a result of their efforts, we are starting to understand how large groups of simple units can interact to produce complex behavior. One of the fruits of this new science has been an understanding of the links between power laws and fractal geometry. Power laws run through biology—in metabolism and other allometries, and in patterns of species diversity, rarity, and commonness. The laws are equally ubiquitous in the physical processes that shape life's environment. Many theories of biology and physics work at some scales but not others—but power laws show how the same principles can apply across scales from mitochondria to sequoias. Scientists have only just begun to exploit fractal geometry and power laws as ways of describing and unifying nature.

The Other Pillar

Attempts at deriving general laws for biology have always met with skepticism and controversy. Often this skepticism proves justified. Most of the patterns proposed as general laws of nature—such as Rubner's surface law, Pearl's rate-of-life hypothesis, Kyoji Yoda's self-thinning in plants, and Evelyn Hutchinson's ratios—have turned out to be not as general, or as accurate, as their advocates first thought. Metabolic ecology is science in the making, not received wisdom like the laws of natural selection, motion, and thermodynamics. The current state of the theory is probably not the last word on the matter, but then nothing in science ever is. "People end up being well known more for the questions they ask than for the answers they provide," comments David Tilman. "West, Brown, and Enquist are asking the right questions."

West, Brown, and Enquist are not the first of biology's unifiers to focus on energy. It's obvious why energy is a good place to seek generality: The chemistry of metabolism is universal, as is the value of energy to organisms. We cannot guarantee that evolution will make organisms faster, or smarter, or longer lived, but it will always reward those that can grab energy and turn it into copies of themselves. So this book has followed the trail of researchers investigating metabolic rate, from a calculation of how much to feed French cigar makers to a hypothesis about the rate at which new species come into being and spread themselves over the globe.

How far could a focus on energy take us? Jim Brown thinks of life as resting on two pillars—energy and information. His work has focused on energy, in the form of metabolism, and he has sought to show how the large-scale structure of nature depends on how individuals get and use energy. There is already a theory dealing with how life uses information—genetics, which explains how the information in DNA is used to make living things. Biology is the product of the interaction between energy and information, similar to the way that a computer's performance is the product of both its hardware and its software. How metabolism influences genetics, and vice versa, is not understood, but there are some hints.

Creatures with more DNA, for example, have larger cells and their cells have slower metabolic rates. This has nothing to do with genes—most DNA (95 percent in humans) seems to be hitchhiking, performing no function in the organism that carries it. But cell size scales positively with the brute quantity of DNA, so creatures with lots of DNA, such as salamanders, have large cells and slow metabolic rates. Birds, on the other hand, have relatively small quantities of DNA and fast metabolic rates. Cells of large animals, which are burning fuel more slowly, also have less RNA, the chemical that life uses to translate genomic information into proteins, just as they have fewer mitochondria. The size and workings of a library depend on how many books it has, not what's written in the books. Similarly, the amount of information a cell needs to handle seems to influence its physical structure and to place limits on the way it uses energy.

There is lots that metabolic ecology will never explain, such as the structure of food webs, why species are common and rare, and why they live in some places and not others. Other ecologists are working to find the general rules that might govern these patterns, in the hope that in a decade or two there will be a set of theories that can explain the patterns in nature. This begs the question of whether such theories could be linked into something even more general, something ecologists have already begun to explore. Drew Allen is combining metabolic theory and neutral theory to investigate the dynamics of evolution and extinction in foraminifera. Brian Enquist is teaming up with ecologists who study food webs and with those who study the patterns in how widespread and how common species are, to try and work out how size influences feeding behavior, and how animals use space. Steve Hubbell is working with physicists to incorporate energy into neutral ecology and to get a handle on species diversity in the process. Perhaps, he thinks, a metabolic view of mutation rates can help quantify the rate at which new species evolve, and provide a variable to plug into the formula for the universal biodiversity number. One day, Hubbell expects there to be a theoretical prediction of the number of species on Earth. Geoff West is dreaming of a grand unified theory of ecology that would bring all these ideas together, just as physicists seek links between quantum theory and relativity. Across the board, macroecologists are seeing where their ideas can take them, forming new collaborations, and creating a few antagonisms.

The Silent Majority

But we should also remember how little we know about nature. We have firm notions about how life works. It would be shocking—at least as momentous as finding extraterrestrial life—to discover a bacterium that did not use DNA to transmit genetic information or that had a genetic code radically different from the one humans use, or that didn't use ATP as an all-purpose molecular fuel. It would be almost as surprising to find a mammal as small as a bee or an insect the size of a chicken. Yet we have almost no idea whether there are 5 million, 50 million, or 500 million species in the world. There are surprises to be had on even the best-known branches of the tree of life. The day I wrote this passage a new species of African monkey, the highland mangabey, was announced. But, while we know a lot about primates—and mammals in general—there are many organisms about whose diversity we know very little. Biologists have described about 100,000 species of fungus and believe there are about 1.5 million in the world. There are thought to be a similar number of nematode worm species living in the sea, but only a few have been named. Estimates of the number of insect species vary hugely, but most biologists currently plump for about 5 million to 10 million.

But it is in the world of the cells called prokaryotes—the species without mitochondria and other complex cellular organs—that things really start to boggle the mind. Until recently, there was no real way to get a handle on microbial diversity. Biologists could only study the microbes they could culture in the lab, but more than 99 percent of bacterial species will not breed in captivity, making them impossible to identify. DNA technology has given us a new way to identify microbes. We can fish bits of DNA out of the soil, sea, sewage, or wherever and amplify and sequence them. By looking at genes, biologists could get some idea of what was out there. The results have been challenging.

There are, it seems, about 70 species of bacteria in a milliliter of sewage. This number rises to 500 species in the human gut, with the mouth, skin, and genitals all having their own distinctive and similarly diverse flora. There could be 2 million bacterial species in the oceans and 4 million in a tonne of soil. The biggest estimate for global bacterial diversity I have seen is 1 billion species. The overwhelming diversity of microbes asserts itself not just at the species level. The kingdom archaea, a group of unicellular organisms that look like regular bacteria but are genetically as different from them as we are, was first discovered living in hotsprings about 25 years ago. Currently, more than one new phylum of prokaryotes—the rank equivalent to molluscs or vertebrates—is being discovered every month.

It's not just in genetic diversity that microbes win. Worldwide, it's been estimated that there are 5×10^{30} (i.e., a 5 followed by 30 zeros) prokaryotes. This is a billion times the number of stars in the universe. These cells contain the same mass of carbon as all the world's plants and 10 times more nitrogen and phosphorus. There are microbes whose metabolism is based on burning not oxygen but uranium, and others that use ammonia. They live in waters heated above boiling point and in streams and stomachs filled with concentrated acid. In terms of diversity, we large organisms can beat them on the variety of shapes we come in and our behavior, but that's it.

Most of what we think we know about ecology, and most patterns and explanations with claims to generality, have come about from studying birds, mammals, and plants. Yet any general theory must also apply to the vast iceberg of biodiversity that we cannot see. Some theories, such as Kleiber's rule, seem to do so—although this is still disputed. And Brian Enquist's team has found that patterns in the population density of marine plankton match the predictions of John Damuth's energy equivalence rule; that is, the number of individuals in an area declines proportional to the $-3/4$ power of their body size.

But microbes will present stiff challenges to any attempt to find general rules that explain diversity. For starters, belonging to a species is not the same thing for microbes as it is for larger organisms. Most microscopic organisms reproduce without sex, by simple cell division. Even more confusing, they can have sex without reproducing. Bacteria can swap DNA with distantly related fellows or even pick up stray bits of DNA from their environment and incorporate them into their genomes. We will need new definitions, and new measures, of diversity. For example, a common criterion for assigning two microbes to different species is whether one particular stretch of their DNA is less than 97 percent alike. For animals this would lump all the primates from humans to lemurs into the same group.

Microbial ecology, and the patterns in microbial diversity, might not fit the theories devised for plants and animals. Large animals and plants are stuck in a certain place—wildebeest in Africa, mahogany in South America. But microbes have such huge populations and can travel so easily, on air and water currents, that some ecologists think any species can get to wherever conditions are right for it, so any saltmarsh will eventually contain all the world's species that can live in salty, anoxic mud, and any hotspring will contain all the species that like hot, sulfurous water. It's as if there were polar bears in Antarctica and penguins in Alaska.

Some microbial ecologists are now challenging the view that everything is everywhere, arguing that the genetic evidence is revealing patterns in the microbial life of hotsprings and saltmarshes that are similar to those seen in the trees or birds in a forest. For example, places become less biologically similar the farther apart they are—which might seem obvious but is still contentious. The science, and debate around it, is only just getting under way. It may be that some view of life can bring bacteria and other microbes into line with other species, or it may be that microscopic life needs new rules. Either way, the possibility that we might have to junk our general theories should not dampen excitement about what biologists are discovering about the microbial world. Ecologists seeking generality have only one data point, the earth. This makes testing ideas difficult. But the microbial world gives us somewhere to test a century's worth of theories built by ecologists who have studied plants and animals. It's like having a whole other planet to play with.

11 EPILOGUE: "THE GREAT DESIDERATUM"

S T. ANDREWS DID NOT LOOK gray in the spring sunshine, although it was still cold. The North Sea breeze made it feel more like March than May. D'Arcy Thompson's old house on South Street has a slate plaque outside commemorating his residence: "naturalist, scholar," it calls him. I had come here—neatly but coincidentally, it was almost exactly a year after I'd had my metabolic rate measured—to see the archive of D'Arcy Thompson's papers. I was hoping for a view into his mind, hoping to track the genesis of his ideas. I was also looking for an ending, some document, anecdote, or experience that would wrap things up tidily.

Sure enough, his notebooks cover the gamut—it was like seeing *On Growth and Form* chopped into bits and reassembled at random. There were stirring quotations: "The great desideratum for any science is its reduction to the smallest number of dominating principles." There were notes on how larger animals had longer gestation periods, slower heartbeats, and longer lives. There was a paper comparing the metabolic rate of prawns from English and Arctic waters and a note on how herring eggs hatched quicker at higher temperatures. There were doubts that natural selection could explain the colors of insects or the shapes of leaves and horns. There were doodles of aeroplane and dragonfly wings, honeycomb, cell shapes, and bone structures. There were lists of numbers and sketched graphs and a correspondence with A. C. Aitken at the University of Edinburgh over some data D'Arcy had collected on the allometry of the stag beetle's body parts. ("What is interesting in your sample is the *disproportionate* growth at different ranges of size—the larger beetle is not a magnification of the smaller one," Aitken wrote.) There was speculation on diversity, from one of D'Arcy's book reviews: "We begin by marvelling at the wealth of species in the world, and end by realizing that, in contrast to infinite variety, they are very few." There was a letter of warning from Marcus Hartog, a leading cell biologist of the late nineteenth century: "I would advise you to be on your guard against the physicists: they love to simplify the problem out of all real relation to the facts of the organism."

Connections emerged; degrees of separation shrank. A quote from Alexander von Humboldt on the physiology of electric fish had been typed up and pasted into a notebook. Evelyn Hutchinson's review of the second edition of *On Growth and Form* was stuck into Thompson's own copy ("Sir D'Arcy still seems to us fresh from conversation with Aristotle, and still seems to have discussed his material with Galileo"). The editor of *Nature* had written, asking Thompson to adjudicate on whether a recent paper published in the journal by the Italian mathematician Vito Volterra had reproduced work already published by Alfred Lotka: "Volterra hasn't got a leg to stand on," D'Arcy concluded. The theorem, which models the effect that predators have on the numbers of their prey, and vice versa, is now called the Lotka-Volterra equation. A paper cut out of a 1936 edition of *Nature* suggested that Costa Rica's plant diversity was due to its high mountains exposing species to mutation-causing cosmic rays.

But yet again, the diversity of ideas and information threatened to become unmanageable. There were notes on the size of nitrogen atoms and comparisons of the size and density of the sun and Betelgeuse; on liquid crystals and molecular films; papers analyzing the relationship between a violin's wood and its acoustic properties; comparisons between the shapes of water lilies and classical vases; a paper on meteorite craters; another on the mathematical psychology of war, giving equations claiming to calculate the likelihood of conflict; a letter cut out of a newspaper on how much start one would need to outrun a charging elephant (35 yards, apparently the elephant gives up after 200); another cutting on the largest haddock ever landed at Aberdeen; a pamphlet titled "Was Dickens What Is Called a Gentleman?"; and a paper on the musical scale of the highland bagpipe.

As I sat in the library, leafing through scrapbooks and piles of correspondence, trying to decipher Thompson's handwriting—which is just about legible when he is writing to someone else and all but impenetrable when he is making notes for personal consumption, especially as he switches between English, German, French, Latin, and Greek—I realized that I was daft to be hoping to find a thank-you-and-goodnight moment in the jottings of a man who missed his deadlines by decades and whose final work was 10 times longer than originally intended.

But, I realized, I was also daft to be looking for a neat ending at all. D'Arcy's notebooks lift the lid on the stew of ideas, arguments, scenic diversions, lucky accidents, and happy mistakes that were boiled down into *On Growth and Form*. A similar blend lies behind all scientific work, be it a book, academic paper, or mathematical model. The finished result might look like a tale of serene progress, but it is also an attempt to turn a deaf ear to the clamor of everything left out—all those intriguing, troubling, and inconvenient things trying to force their way back into the picture. Whether people see unity or diversity, solubility or impossibility, says something about them as well as something about the world, and if you choose a different path or a different viewpoint, you will see something different. Why should I expect one conclusion, when every idea is still so contentious and every question throws up so many possible answers?

This is not an appeal to relativism, to the idea that all explanations are equally good. The unity of nature might turn out to be a mythical beast, but those hunting it will probably find useful things along the way. If a handful of rules—based, say, on metabolism, genetics, and statistics—can explain much of how life works, from molecules to oceans, from forams to quetzals, and explain why life shows both such diversity and such pattern, it will be a stunning achievement. But it won't be the end. The theories will be there to be challenged, improved, and replaced, and the diversity will still be there to beguile and puzzle. Is nature beautifully simple or beautifully complex? Yes, it is.

ACKNOWLEDGMENTS

MY BIGGEST DEBT of thanks is to Brian Enquist and his team, Brad Boyle, Drew Kerkhoff, Jason Pither, and Nate Svensson, for having me along on their field trip to Costa Rica, for being welcoming, and for their patience in explaining and showing how ecology is done. Thanks also to Jim Brown, Jamie Gillooly, and Geoff West for their time in New Mexico. Jim Brown and Brian Enquist also read a draft manuscript and supplied helpful comments and corrections, as did John Damuth, of the University of California, Santa Barbara; Lev Ginzburg of the State University of New York, Stony Brook; Patty Gowaty at the University of Georgia, Athens; and Jessica Green of the University of California, Merced.

Thanks are also due to all the researchers who took time to answer my questions, explain concepts, share their opinions, send me their papers, and tell me their stories, both for this book and for earlier pieces that informed the book. In alphabetical order they are Drew Allen, National Center for Ecological Analysis and Synthesis, Santa Barbara; Jayanth Banavar, Pennsylvania State University; Folmer Bokma, University of Groningen; Chris Carbone, Institute of Zoology, London; Peter Chesson, University of California, Davis; Robert Colwell, University of Connecticut; Vasiliki Costarelli, South Bank University; Peter Dodds, Columbia University; Stephen Gibson, University of Sydney; Alfred Heusner, University of California, Davis; Steven Heymsfield, St. Luke's Roosevelt Hospital, New York; Michael Huston, Texas State University; Walter Jetz, University of California, San Diego; Jan Kozlowski, Jagiellonian University, Cracow; Neo Martinez, Pacific Ecoinformatics and Computational Ecology Lab; Robert May, University of Oxford; Kevin McCann, University of Guelph; Brian McNab, University of Florida; Shai Meiri, Imperial College, London; Jim Miller, Missouri Botanical Garden; Bertram Murray, Rutgers University; Sean Nee, University of Edinburgh; Karl Niklas, Cornell University; Andrew Numa, South Eastern Sydney Area Health Service; Han Olff, University of Groningen; Stuart Pimm, Duke University; Andy Purvis, Imperial College, London; John Reynolds, University of East Anglia; Robert Ricklefs, University of Missouri, St. Louis; Mark Ritchie, Syracuse University; David Robinson, University of Aberdeen; Klaus Rohde, University of New England; Michael Rosenzweig, University of Arizona; Van Savage, Harvard University; Nancy Slack; Lawrence Slobodkin, State University of New York, Stony Brook; John Speakman, University of Aberdeen; and Wayne van Voorhies, Arizona State University. Norman Reid and his colleagues at St. Andrews University Library greatly aided my research in the D'Arcy Thompson archive, and Alison Hopkins provided invaluable help with illustrations. Any mistakes are my fault.

Thanks also to my agent, Peter Tallack, for selling this book and my editor, Jeffrey Robbins, for buying it and for his work on the finished product. And especially to my mum and dad for their support.

Web Site

The following reference list is skeletal, containing only the most relevant, recent, and (physically, electronically, and intellectually) accessible sources. There is a longer and ongoing list of references relating to the topics in this book online at *www.inthebeatofaheart.com*. The site also contains extra figures and illustrations, updates on the science described in this book, a blog, and corrections.

REFERENCES

General References

Blackburn, T. M., and K. J. Gaston. 2003. *Macroecology: Concepts and Consequences.* (Blackwell, Oxford).

Brown, J. H. 1995. *Macroecology* (University of Chicago Press, Chicago, IL).

Calder, W. A., III. 1996. *Size, Function and Life History* (Dover, Mineola, NY).

Kingsland, S. 1995. *Modeling Nature: Episodes in the History of Population Ecology,* 2nd edition (University of Chicago Press, Chicago, IL).

Peters, R. H. 1983. *The Ecological Implications of Body Size* (Cambridge University Press, Cambridge).

Rosenzweig, M. L. 1995. *Species Diversity in Space and Time* (Cambridge University Press, Cambridge).

Schmidt-Nielsen, K. 1984. *Scaling: Why Is Animal Size so Important?* (Cambridge University Press, Cambridge).

Thompson, D. W. 1961. *On Growth and Form.* Abridged edition, edited by J. T. Bonner, foreword by S. J. Gould (Cambridge University Press, Cambridge).

1 Prologue: "I Have Taken to Mathematics"

Fox Keller, E. 2002. *Making Sense of Life: Explaining Biological Development with Models, Metaphors, and Machines* (Harvard University Press, Cambridge, MA).

Gould, S. J. 1971. D'Arcy Thompson and the science of form. *New Literary History* 2:229–258.

Thompson, D. W. 1940. *Science and the Classics* (Oxford University Press, Oxford).

Thompson, D. W. 1942. *On Growth and Form*, 2nd edition (Cambridge University Press, Cambridge).

Thompson, R. D. 1958. *D'Arcy Wentworth Thompson, the Scholar-Naturalist*. With an afterword by Peter Medawar (Oxford University Press, Oxford).

2 The Slow Fire

Brody, S. 1945. *Bioenergetics and Growth* (Rheinhold, New York, NY).

DuBois, E. F. 1936. *Basal Metabolism in Health and Disease*, 3rd edition (Baillière, Tindall and Cox, London).

Ferguson, S. H., and S. Larivière. 2004. Are long penis bones an adaptation to high latitude snowy environments? *Oikos* 105:255–267.

Garrow, J. S., W. P. T. James, and A. Ralph. 1999. *Human Nutrition and Dietetics*, 10th edition (Churchill Livingstone, London).

Haldane, J. B. S. 1985. *On Being the Right Size and Other Essays* (Oxford University Press, Oxford).

McNab, B. K. 1992. Energy expenditure: A short history. Pp. 1–15 in *Mammalian Energetics*, edited by T. E. Tomasi and T. H. Horton. (Cornell University Press, Ithaca, NY).

Meiri, S., and T. Dayan. 2003. On the validity of Bergmann's rule. *Journal of Biogeography* 30:331–351.

Rothschuh, K. E. 1975. Rubner, Max. *Dictionary of Scientific Biography*, edited by C. C. Gillespie, vol. X1, pp. 585–586 (Scribner's, New York, NY).

Rubner, M. 1982. *The Laws of Energy Consumption in Nutrition*. Translated by A. Markoff and A. Sandri-White, edited by Robert J. T. Joy, with a biography of Max Rubner by William H. Chambers (Academic, London).

3 Moving the Line

Benedict, F. G. 1938. *Vital Energetics: A Study in Comparative Basal Metabolism* (Carnegie Institute, Washington, DC).

Clark, R. W. 1960. *Sir Julian Huxley* (Phoenix House, London).

Gibson, S., and A. Numa. 2003. The importance of metabolic rate and the folly of body surface area calculations. *Anaesthesia* 58:50–83.

Huxley, J. 1972. *Memories* (Penguin, London).

Huxley, J. 1972. *Problems of Relative Growth* (Dover, New York, NY).

Kleiber, M. 1932. Body size and metabolism. *Hilgardia* 6:315–353.

Kleiber, M. 1961. *The Fire of Life: An Introduction to Animal Energetics* (Wiley, New York, NY).

4 Searching for Similarity

The best publications to start reading about the hypotheses advanced to explain the relationship between body mass and metabolic rate up to the mid-1980s are Peters's *The Ecological Implications of Body Size*, and Schmidt-Nielsen's *Scaling*.

Farley, C. T. 1999. Thomas A. McMahon (1943–99). *Nature* 398:566.

Heusner, A. A. 1991. Size and power in mammals. *Journal of Experimental Biology* 160:25–54.

McMahon, T. 1973. Size and shape in biology. *Science* 179:1201–1204.

Plaut, K., et al. 2003. Effects of hypergravity on mammary metabolic function: Gravity acts as a continuum. *Journal of Applied Physiology* 95:2350–2354.

Wang, Z., et al. 2001. The reconstruction of Kleiber's law at the organ-tissue level. *Journal of Nutrition* 131:2967–2970.

West, L. J., C. M. Pierce, and W. D. Thomas. 1963. Lysergic acid diethylamide: Its effects on a male Asiatic elephant. *Science* 138:1100–1103.

5 Networking

Banavar, J. R., et al. 2002. Supply-demand balance and metabolic scaling. *Proceedings of the National Academy of Sciences USA* 99:10506–10509.

Calder, W. A. Diversity and convergence: Scaling for conservation. Pp. 297–324 in *Scaling in Biology*, edited by J. H. Brown and G. B. West. (Oxford University Press, Oxford).

Gillooly, J. F. 2001. Effects of size and temperature on metabolic rate. *Science* 293:2248–2251.

Lotka, A. J. 1956. *Elements of Mathematical Biology* (Dover, New York, NY).

West, B. J., and A. L. Goldberger. 1987. Physiology in fractal dimensions. *American Scientist* 75:354–365.

West, G. B., and J. H. Brown. 2004. Life's universal scaling laws. *Physics Today* 57:36–42.

West, G. B., et al. 1997. A general model for the origin of allometric scaling laws in biology. *Science* 276:122–126.

West, G. B., et al. 2000. The origin of universal scaling laws in biology. Pp. 87–112 in *Scaling in Biology*, edited by J. H. Brown and G. B. West. (Oxford University Press, Oxford).

Whitfield, J. 2001. All creatures great and small. *Nature* 413:342–344.

6 The Pace of Life

Austad, S. N. 1997. *Why We Age: What Science Is Discovering About the Body's Journey Through Life* (Wiley, New York, NY).

Bertalanffy, L. v. 1933. *Modern Theories of Development: An Introduction to Theoretical Biology*. Translated and adapted by J. H. Woodger (Oxford University Press, Oxford).

Bertalanffy, L. v. 1952. *Problems of Life: An Evaluation of Modern Biological Thought* (Watts, London).

Brown, J., and G. West. 2004. One rate to rule them all. *New Scientist* (May 1) 38–41.

Gillooly, J. F., et al. 2001. Effects of size and temperature on developmental time. *Nature* 417:70–73.

Guiot, C., et al. 2003. Does tumor growth follow a "universal law"? *Journal of Theoretical Biology* 225:147–151.

McNab, B. K. 2003. Ecology shapes bird bioenergetics. *Nature* 426:620–621.

Moses, M. E., and J. H. Brown. 2003. Allometry of human fertility and energy use. *Ecology Letters* 6:295–300.

Nemoto, S., and T. Finkel. 2004. Ageing and the mystery at Arles. *Nature* 429:149–152.

Pearl, R. 1928. *The Rate of Living, Being an Account of Some Experimental Studies on the Biology of Life Duration* (Knopf, New York, NY).

Speakman, J. R., et al. 2004. Uncoupled and surviving: Individual mice with high metabolism have greater mitochondrial uncoupling and live longer. *Aging Cell* 3:87–95.

Sterner, R. W., and J. J. Elser. 2002. *Ecological Stoichiometry: The Biology of Elements from Molecules to the Biosphere* (Princeton University Press, Princeton, NJ).

West, G. B., et al. 2001. A general model for ontogenetic growth. *Nature* 413:628–631.

Whitfield, J. 2004. Ecology's big, hot idea. *PLoS Biology* 2:2023–2027.

7 Seeing the Forest for the Trees

Baldocchi, D., et al. 2001. FLUXNET: A new tool to study the temporal and spatial variability of ecosystem-scale carbon dioxide, water vapor, and energy flux densities. *Bulletin of the American Meteorological Society* 82:2415–2434.

Carbone, C., and J. L. Gittleman. 2002. A common rule for the scaling of carnivore density. *Science* 295:2273–2276.

Damuth, J. 1987. Interspecific allometry of population density in mammals and other animals: The independence of body mass and population energy use. *Biological Journal of the Linnean Society* 31:193–246.

Enquist, B. J., and K. Niklas. 2001. Invariant scaling relations across tree-dominated communities. *Nature* 410:655–660.

Enquist, B. J., et al. 2003. Scaling metabolism from organisms to ecosystems. *Nature* 423:639–642.

Hutchings, M. 1983. Ecology's law in search of a theory. *New Scientist* (June 16)765–767.

Jetz, W., et al. 2004. The scaling of animal space use. *Science* 306:266–268.

Niklas, K. 1994. *Plant Allometry* (University of Chicago Press, Chicago, IL).

Niklas, K. 2004. Plant allometry: Is there a grand unifying theory? *Biological Reviews* 79:871–889.

Savage, V. M., et al. 2004. Effects of body size and temperature on population growth. *American Naturalist* 163:429–441.

8 The Cult of Santa Rosalia

Chase, J. M., and M. A. Leibold. 2003. *Ecological Niches: Linking Classical and Contemporary Approaches* (University of Chicago Press, Chicago, IL).

Colinvaux, P. C. 1980. *Why Big Fierce Animals Are Rare* (Penguin, London).

Diamond, J. 1975. The assembly of species communities. Pp. 342–444 in *Ecology and Evolution of Communities*, edited by M. L. Cody and J. M. Diamond. (Harvard University Press, Cambridge, MA).

Horn, H. S., and R. M. May. 1977. Limits to similarity among coexisting competitors. *Nature* 270:660–661.

Hubbell, S. P. 2001. *The Unified Neutral Theory of Biodiversity and Biogeography* (Princeton University Press, Princeton, NJ).

Hutchinson, G. E. 1959. Homage to Santa Rosalia, or why are there so many kinds of animals? *American Naturalist* 93:145–159.

Lewin, R. 1983. Santa Rosalia was a goat. *Science* 221:636–639.

Lindeman, R. L. 1942. The trophic-dynamic aspect of ecology. *Ecology* 23:399–418.

MacArthur, R. 1968. The theory of the niche. Pp. 159–176 in *Population Biology and Evolution*, edited by R. C. Lewontin. (Syracuse University Press, Syracuse, NY).

McCann, K. S. 2000. The diversity-stability debate. *Nature* 405:228–233.

Ritchie, M. E., and H. Olff. 2000. Spatial scaling laws yield a synthetic theory of biodiversity. *Nature* 400:557–560.

Silvertown, J. 2005. *Demons in Eden: The Paradox of Plant Diversity* (Chicago University Press, Chicago, IL).

Simberloff, D., and W. Boecklen. 1981. Santa Rosalia reconsidered: Size ratios and competition. *Evolution* 35:1206–1228.

Slobodkin, L. B., and N. G. Slack. 1999. George Evelyn Hutchinson: 20th Century ecologist. *Endeavour* 23:24–30.

Tilman, D. 2000. Causes, consequences and ethics of biodiversity. *Nature* 405:208–211.

Whitfield, J. 2002. Neutrality versus the niche. *Nature* 417:480–481.

Wilson, E. O. 1995. *Naturalist* (Penguin, London).

9 Humboldt's Gifts

Rosenzweig's *Species Diversity in Space and Time* is a good place to start reading about the latitudinal gradient in diversity.

Allen, A. P., et al. 2002. Global biodiversity, biochemical kinetics, and the energetic-equivalence rule. *Science* 297:1545–1548.

Brown, J. H. 1988. Species diversity. Pp. 57–89 in *Analytical Biogeography: An Integrated Approach to the Study of Animal and Plant Distributions*, edited by A. A. Myers and P. S. Giller. (Chapman & Hall, London).

Colwell, R. K., et al. 2004. The mid-domain effect and species richness patterns: What have we learned so far? *American Naturalist* 163:E1–E23.

Emerson, B. C., and N. Kolm. 2005. Species diversity can drive speciation. *Nature* 434:1015–1017.

Gendron, V. 1961. *The Dragon Tree. A Life of Alexander, Baron von Humboldt* (Longmans, Green, New York, NY).

Humboldt, A. v. 1849. *Aspects of Nature in Different Lands and Different Climates* (Longman, London).

Humboldt, A. v. 1995. *Personal Narrative of a Journey to the Equinoctal Regions of the New Continent* (Penguin, London).

Raby, P. 2001. *Alfred Russel Wallace* (Chatto & Windus, London).

Wallace, A. R. 1969. *Natural Selection and Tropical Nature. Essays on Descriptive and Theoretical Biology* (Gregg International, Farnborough, England).

Wallace, A. R. 1980. *My Life* (Chapman & Hall, London).

Willdenow, C. L. 1805. *The Principles of Botany, and of Vegetable Physiology.* Translated from German (Edinburgh, 1805).

Willig, M. R., et al. 2003. Latitudinal gradients of biodiversity: Pattern, process, scale, and synthesis. *Annual Review of Ecology and Systematics* 34:273–309.

10 A Newton of the Grass Blade?

Ginzburg, L., and M. Colyvan. 2004. *Ecological Orbits: How Planets Move and Populations Grow* (Oxford University Press, Oxford).

Gittleman, J. L., and A. Purvis. 1998. Body size and species-richness in carnivores and primates. *Proceedings of the Royal Society London B* 265:113–119.

Kant, I. 1987. *Critique of Judgement* (Hackett, Indianapolis, IN).

Lawton, J. H. 1999. Are there general laws in ecology? *Oikos* 84:179–192.

May, R. M. 1999. Unanswered questions in ecology. *Philosophical Transactions of the Royal Society London B* 354:1951–1959.

Murray, B. G. 2001. Are ecological and evolutionary theories scientific? *Biological Reviews* 76:255–289.

Nee, S. 2004. More than meets the eye. *Nature* 429:804–805.

Simberloff, D. 2004. Community ecology: Is it time to move on? *American Naturalist* 163:787–799.

Slobodkin, L. B. 2001. The good, the bad and the reified. *Evolutionary Ecology Research* 3:1–13.

Thomas, C. D., et al. 2004. Extinction and climate change. *Nature* 427:145–148.

Whitfield, J. 2005. Biogeography: Is everything everywhere? *Science* 310:960–961.

INDEX